J. Kammeyer would like to thank the U.S. Naval Academy, in particular for their sabbatical support during the course of this research.

D. Rudolph gratefully acknowledges the support of NSF grants DMS9401538 and DMS9706829.

The purpose of this work is to lift the notion of restricted orbit equivalence to the category of free and ergodic actions of discrete amenable groups. The axiomatics of a size m and the nature of the associated equivalence relation, m-equivalence, are established. An extensive list of examples of sizes and the corresponding equivalence relations are described. An entropy, called m-entropy, associated with each size, is defined as the infimum of the classical entropy over the m-equivalence class. It is proven that a restricted orbit equivalence is either entropy preserving, in that the m-entropy is simply the classical entropy, or entropy free, in that on a residual subset of the equivalence class, the entropy is zero and hence the m-entropy of all actions is zero. The notion of m-finitely determined is introduced, and some of its basic properties are developed, in particular that it is an m-equivalence invariant. Finally, the equivalence theorem is proven, that any two m-finitely determined actions of equal m-entropy are m-equivalent. This is carried out using category methods, following the Burton–Rothstein approach, within a natural Polish space of m-joinings. In the Appendix, it is shown that previous size axiomatizations give rise to essentially the same notion of m-equivalence.

Restricted Orbit Equivalence for Actions of Discrete Amenable Groups

Janet Whalen Kammeyer

Daniel J. Rudolph

CAMBRIDGE
UNIVERSITY PRESS

CAMBRIDGE UNIVERSITY PRESS
Cambridge, New York, Melbourne, Madrid, Cape Town, Singapore,
São Paulo, Delhi, Dubai, Tokyo, Mexico City

Cambridge University Press
The Edinburgh Building, Cambridge CB2 8RU, UK

Published in the United States of America by Cambridge University Press, New York

www.cambridge.org
Information on this title: www.cambridge.org/9780521183857

First published 2002
First paperback edition 2010

A catalogue record for this publication is available from the British Library

Library of Congress Cataloguing in Publication data

Kammeyer, Janet Whalen, 1963–
Restricted orbit equivalence for actions of discrete amenable groups / Janet Whalen Kammeyer, Daniel J. Rudolph.
p. cm
Includes bibliographical references and index.
ISBN 0 521 80795 6
1. Measure-preserving transformations. 2. Entropy (Information theory)
I. Rudolph, Daniel J. II. Title.

QA613.7.K36 2001
515′.42–dc21 2001025506

ISBN 978-0-521-80795-1 Hardback
ISBN 978-0-521-18385-7 Paperback

Contents

1
Introduction

1.1 Overview

The purpose of this work is to lift the notion of restricted orbit equivalence to the category of free and ergodic actions of discrete amenable groups. We mean *lift* in two senses. First of course we will generalize the results in [43], where Rudolph developed a theory of restricted orbit equivalence for $\mathbb{Z}$-actions, and in [25] where both authors later established a similar theory for actions of $\mathbb{Z}^d, d \geq 1$ to actions of these more general groups. However, we will also *lift* in the sense that we will develop the axiomatics and argument structures in what we feel is a far more natural and robust fashion. Both [43] and [25] were based on axiomatizations of a notion called a "size" measuring the degree of distortion of a box in $\mathbb{Z}^d$ caused by a permutation. It is not evident that on their common ground, $\mathbb{Z}$-actions, these two theories agree. Hence we refer to the first as a 1-size and the second as a p-size, p for "permutation".

Here we will establish the axiomatics of what we will simply call a size. We ask that the reader accept this new definition. In the Appendix we show that any equivalence relation that arose from a p-size will arise from a size as we define it here. The same is not done for 1-sizes, but for a slight strengthening of this axiomatics that includes all the examples in [43].

We will work on the level of countable and discrete amenable groups, as the work of Ornstein and Weiss [37] has shown this to be a natural level on which all the basic dynamics and ergodic theory we need holds sway. We will not push beyond this to locally-compact amenable groups as the formalism of both orbit equivalence and entropy theory require basic work before our approach appears feasible.

Outlining our approach, in Section 2.1 we will set up the basic vocabulary we will use for our work, the vocabulary of arrangements and rearrangements of orbits, and describe the natural topologies on these spaces. In Section 2.2 we will establish the axiomatics of a size *m* and the nature of the associated equivalence relation, *m*-equivalence. One sees immediately here the change in perspective from [43] and [25] in that a size *m* is now a family of pseudometrics on the full-group of a free and ergodic action, one for each arrangement of the orbit as an action of the group *G*. The *m*-equivalence class of an arrangement will appear here as a certain G_δ subset of the completion of the full-group relative to this pseudometric. We end Chapter 2 with a list of seven equivalence relations, some well known others not so well known, which can be described as *m*-equivalences for an appropriate *m*. We also present one "non-example" that uses the methods developed here but does not fall directly under our development and indicates one of several directions in which to further broaden this approach.

In Chapter 3 we present the fundamental results that we will need from the Ornstein and Weiss work on the ergodic theory of actions of amenable groups. In Chapter 4 we present a variety of copying lemmas that will be essential to our progress both in developing an entropy theory for restricted orbit equivalences and for our proof of the equivalence theorem. Chapter 5 contains our development of an entropy theory for restricted orbit equivalences. We define an entropy, called *m*-entropy, associated with each size as the infimum of the classical entropy on the *m*-equivalence class. The principle result we obtain, (as was done in [43] and [25] for the cases they considered) is that a restricted orbit equivalence is either entropy-preserving, in that *m*-entropy is simply the classical entropy, or entropy-free in that on a residual subset of the equivalence class the entropy is zero and hence the *m*-entropy of all actions is zero.

From this point our goal is to prove the natural generalization of Ornstein's isomorphism theorem for Bernoulli shifts for our restricted orbit equivalences. That is to say, we wish to show that there are certain distinguished free and ergodic actions, intrinsically recognizable, for which *m*-entropy is a complete invariant of *m*-equivalence. As we indicated earlier our goal in *lifting* results is both to demonstrate that they hold in the more general context and also to raise the general level of argument to a more robust form. In particular the approach to the equivalence theorem we take is to bring to bear the categorical approach that Burton and Rothstein [42] brought to the isomorphism theorem.

To accomplish this it is necessary to build up a certain topological perspective. We will be considering Polish spaces and Polish actions. Recall that these are topological spaces that can be imbedded as a G_δ subset of a compact metric space, and homeomorphisms of them. Chapter 6 presents this development from its foundations through to the proof that the space of m-joinings of free and ergodic G-actions form a Polish space of measures. Although we cannot expect the reader to have any idea at this point what precisely an m-joining is, the point of such a result should be clear to the reader familiar with the Burton–Rothstein approach. Any m-equivalence between two actions will sit as a subset of this space of m-joinings. Just as Burton and Rothstein show that for Bernoulli actions of equal entropy, the conjugacies are a residual subset of the space of joinings, our aim is to show that for any two m-Bernoulli actions of equal m-entropy, the m-equivalences sit as a residual subset of the space of m-joinings.

To obtain this in Section 7.2 we introduce the notion of an m-finitely determined action and develop some of its basic properties, in particular that it is an m-equivalence invariant and is inherited by factor actions. What remains to complete the equivalence theorem is to define a list of open sets in the space of m-joinings whose intersection is precisely the m-equivalences, and to show that in the case of m-finitely determined actions they are dense. The first part of this is easy. It is the second part that takes some work. Central to this proof of density is of course the copying lemmas which we developed in Chapter 4. With them we show how to *perturb* an arbitrary m-joining to lie in one of our open sets. Section 7.1 presents the basic structure theory for the notion of *perturbation* we will use. Section 7.3 finally completes the equivalence theorem.

To connect this work with earlier work, in the Appendix we will demonstrate that p-sizes give sizes in our sense with the same equivalence classes, and that 1-sizes essentially do in that all known examples do, and in general the m-finitely determined classes are the same. This means the large classes of examples discussed there are restricted orbit equivalences according to the definition used here.

1.2 A roadmap to the text

We offer the readers an indication of how they might best benefit from this text depending on the level and nature of their interest. This text is not intended as an introduction to the isomorphism theory of Ornstein. A reasonable preparation is necessary before the material presented here will be comfortably accessible. To the reader interested in a broad overview of dynamics we recommend the texts of Walters [62], Petersen [39], Cornfeld, Fomin and Sinai [7], Hasselblatt and Katok [17], and Brin and Stuck [4]. We also recommend the reader consult Ornstein [35], Shields [50], Rudolph [44], and Ornstein, Rudolph and Weiss [38]. Of these [44] is perhaps the most important reference as it is intended as a basic technical introduction to fundamentals at work here. For the reader seeking a deep understanding of our presentation these older works will provide valuable perspective and background.

We envision two audiences for this text, those seeking a basic technical overview of the tools and methods of this theory for use in their own work but without an intention to work in this area and students and researchers seeking a deep understanding of this area with the intention of working in it. We give here an abbreviated path through the text for the former audience and a recommended first reading for the latter.

All readers should spend time on Chapter 2. For a reader only interested in what the text is about, Chapter 2 offers a sufficient treatment. The material of Chapter 3, through Theorem 3.0.6, is the now classical treatment of the ergodic theory of discrete amenable groups due to Ornstein and Weiss [37]. Understanding this is a must for anyone interested in modern ergodic theory. The next few pages introduce the vocabulary of names for the entropy theory of these actions and it is important to understand them. Beyond this, the important conclusions of Chapter 3 are Theorem 3.0.9 and Corollary 3.0.10 and the reader should be familiar with their meaning.

Chapter 4 is quite technical. What is essential are the two Theorems 4.0.5 and 4.0.13 and one should understand their meaning. The technicalities of proof can be absorbed later if needed. What is essential from Chapter 5 concerning m-entropy is contained in its first paragraph.

Chapters 6 and 7 are really only appropriate for the reader wanting a detailed understanding of the equivalence theory. Chapter 6 develops the space of m-joinings of two actions as a Polish space of measures on a symbolic representation. This provides a framework on which the equivalence theorem of Chapter 7 can be reached without much ado. Chapter 6 though is quite heavy going. The reader should read Section 6.1 as an introduction to Polish spaces ending with a brief and vague description of the succession of spaces constructed to reach the notion of an m-joining. From here one can proceed to Chapter 7, skipping over Section 7.1 and instead focusing on the definitions and results of Section 7.2. Continue with Section 7.3 through the statement of Theorem 7.3.3.

The path just described gives a complete overview of the text. To continue the reader should now return for a careful reading of Chapter 6 and then Section 7.1.

1.3 History and references

The notion of restricted orbit equivalence can be traced back some decades, and the techniques used here can be traced back somewhat further. We take this opportunity to outline our understanding of the history of these ideas and to acknowledge the sources of our work. There are many significant parts of this broad area that we will not mention here. We focus on those particular ideas central to the evolution of this particular part of the theory, the construction of general orbit equivalence theorems, and the tools of those constructions.

Certainly the first pieces, historically, of this story do not relate directly to notions of orbit equivalence, but rather are the basic technical point-set tools of ergodic theory, in particular the Rokhlin lemma [31], [41], the mean ergodic theorem [61], and the entropy theory of Kolmogorow and Sinai [28], [29], [51] (we cite the earliest references we know to the basic results and methods). Of these three the Rokhlin lemma may seem the most mundane but in the long run it is in fact the deepest of the three. Looking forward to the seminal work of Ornstein and Weiss [37] on actions of amenable groups, it is their ability to realize a Rokhlin lemma that in fact carries forward their entire program.

Dye's proof, in 1959 [10], that any two ergodic measure preserving actions on non-atomic standard probability spaces are orbit-equivalent is the first instance of an orbit equivalence theorem, and of course a very startling one. The core technical pieces of this proof are of course the Rokhlin lemma and the ergodic theorem. Dye's original techniques are still visible in all the copying lemmas that follow. The only issue that he did not have to address was entropy. Dye's result was simply too profound and final in the case of measure-preserving actions and so has had much more impact in non-singular dynamics and Von Neuman algebras than in ergodic theory per se. Krieger [30], [54] was able to characterize the orbit equivalence classes of all non-singular ergodic actions by careful study of the obstacles to Dye's argument. The obstacle to moving our work in the direction of non-singular dynamics is the issue of entropy. Certainly, though, one might consider developing the notion of entropy-free restricted orbit equivalences in this setting. On the other hand, perhaps the attempt to lift these methods to the non-singular category will give some insight into how to approach entropy in this category as entropy in the measure-preserving category can be defined by a restricted orbit equivalence (see Section 2.3).

The next step, and truly the pivotal one, in this development was Ornstein's proof that any two Bernoulli shifts of equal entropy were isomorphic [32], [33], [34]. One must remember Sinai [52] had already shown they were weakly isomorphic, that is to say each sat as a factor action of the other. Ornstein's real contribution, from our perspective, is his general characterization of the Bernoulli shifts via the notion of finitely determined actions and more generally the wide range of powerful constructive tools he laid out. At this time the notion of finitely determined was given in a finitistic form, not in terms of joinings as is more common now; but the lift to the joinings perspective is a modest contribution in comparison to the significance of the original concept. This theorem is a phenomenal piece of technical work. Others have had profound insights since then but this result showed what the path to an equivalence theorem would look like. All one had to do was see how to take the steps.

One also should note that the Ornstein machinery provides tools for showing not only that Bernoulli shifts of equal entropy are conjugate, but also that various collections of actions are not conjugate. Parallel to the positive side of equivalence/isomorphism results, one could use these methods to develop non-isomorphism results. For example, Ornstein and Shields [36] constructed uncountable families of non-conjugate ergodic K-systems, all of the same entropy. Relating back to Sinai's theorem, Polit [40] produced a pair of weakly-isomorphic but non-isomorphic actions (in this case zero-entropy mixing actions). This side of ergodic theory is extensive.

During the same period Vershik [60] began an investigation building on Dye's work, but taking a different focus. In orbit equivalence terms he was considering actions of groups other than $\mathbb{Z}$, in particular actions of infinite sums of finite cyclic groups. Such a group is the union of a sequence of finite groups $\mathscr{H}_n$, where each $\mathscr{H}_n/\mathscr{H}_{n+1}$ is a cyclic group of order r_n. The simplest non-trivial case is where the r_n's are all two, the dyadic case. What he considered was in fact a notion of restricted orbit equivalence, asking that the orbit equivalence between two such actions should be an orbit equivalence of each of the $\mathscr{H}_n$-subgroups. Heicklen has completely translated this work into the vocabulary of restricted orbit equivalence as we develop it here (see [19], [18] and Example 6 in Section 2.3). In truth this is more naturally described in terms of the $\mathscr{H}_n$-invariant σ-algebras $\mathscr{F}_n$. Notice that these form a decreasing sequence (or reverse filtration) of algebras $\mathscr{F}_{n+1} \subseteq \mathscr{F}_n$. Ergodicity of the action is equivalent to the algebras intersecting to the trivial algebra. Vershik's notion of equivalence is simply that there should be a map between the measure spaces respecting the two reverse filtrations. In this sense Vershik's study has a similar feel to that of Dye, the natural generalizations move away from measure-preserving actions. One is led to consider general reverse filtrations, without regard to their arising from a measure-preserving action.

In terms of orbit equivalence theory though, this particular development has a particularly prescient nature. Vershik showed that there was an entropy associated with this relation and, that if the r_n's did not grow too quickly, it was the standard entropy of the action, and if they grew quickly enough it would be zero. Using this observation both he and Stepin [53] were able to construct non-equivalent reverse filtrations as they had distinct entropies. Central to Vershik's study is the notion of standardness. This also points to a central aspect of later work, and certainly ours, that there will be certain distinguished classes, the "Bernoulli" class of the given equivalence relation. (Vershik's "standard" class is the zero-entropy m-finitely determined class for the associated size m.) This work gives the first clear indication of how entropy might enter in a general picture.

The next major contribution in the direction of our work also had two sides, one in the west and one in the east. The notion of Kakutani equivalence had arisen some decades earlier in the study of measurable cross-sections of measure-preserving actions of $\mathbb{R}$ [2], [3], [22]. It was known that although entropy was not an invariant of Kakutani equivalence, entropy class was (zero, finite, or infinite entropy) and that entropy changed in a simple way when moving among equivalent actions (Abramov's formula [1]). What Feldman did in the west and what Katok did in the former Soviet Union was to introduce the $\bar{f}$-metric (Feldman's notation) on names. Feldman used this to show that there were many distinct Kakutani equivalence classes of the same entropy class [11]. He also saw the possibility that $\bar{f}$ might plug into Ornstein's isomorphism machinery and lead to an equivalence theory parallel to Ornstein's conjugacy theory. Katok [26], in part jointly with Sataev [49], and Ornstein, Weiss and Rudolph fulfilled that expectation [38]. Katok and Sataev, of course, were working completely independently of Feldman, Ornstein and Weiss.

Building on Feldman's original examples, and Ornstein's observation that the Cartesian square of a rank-1 and mixing map would not be finitely fixed, many exotic examples were constructed. This constructive side was also pursued by Katok, leading to his construction, via this theory, of the first smooth K and not Bernoulli action [26].

Building on the existence of three theorems, Dye's, Ornstein's, and the Feldman, Katok, Ornstein, Weiss–Kakutani equivalence theorem, Feldman proposed in 1975 the potential for a general theory of equivalence relations based on the common structures in these three results. There were two essential gaps in the picture. Kakutani equivalence is not a restricted orbit equivalence in that it is not an orbit equivalence. This seemed a minor issue in that inducing on a subset is not so far from an orbit equivalence. More critical though was to understand what role entropy would play. The isomorphism theorem and Kakutani equivalence theorem both use entropy in much the same way, following Ornstein's basic plan. But entropy does not enter Dye's theorem at all. Of course Vershik's work had already indicated what the answer might be, but this was not well known in the west at the time. The one real gap in the picture was exactly how to phrase a general theorem, although the thought certainly was to create the needed material to apply Ornstein's method, that is to say, define some analogue of $\bar{d}$ or $\bar{f}$.

On a more technical level, in the late 1970s Burton and Rothstein gave an approach to Ornstein's proof that recast the focus to joinings [42], [43]. In these terms what one sees is not the detailed construction of a single conjugacy, but rather the description of all conjugacies as a residual subset of a space of joinings. Although in a pure sense there was nothing really new in what Rothstein did, it made the whole path to the result much clearer. The critical technical piece of the isomorphism theorem was the copying lemma which forced the denseness of certain open sets in a space of joinings. The intersection of these sets were precisely the factor maps, or projections, of some fixed but arbitrary system of sufficient entropy onto a finitely determined system. In this light one could say that Sinai had it right, that although weak isomorphism did not imply isomorphism, if the class of projections of a general system on a finitely determined system of equal entropy could be shown large enough (that is, shown residual in the compact space of all joinings) then there would be lots of isomorphisms to choose from between two finitely determined systems of equal entropy and residual set of them. It was not clear until our work here that this perspective could be lifted to the more general restricted orbit equivalence level. In particular [25], [37] and [43] all follow the original direct constructive approach of Ornstein.

When del Junco and Rudolph showed that Kakutani equivalence could be given a natural characterization in terms of orbit equivalence [9] the general picture became much clearer. As we pointed out earlier, from Abramov's formula we know that the entropy of an induced map varies inversely proportionally to the measure of the set on which one induces. If one defines a notion of "even equivalence" of two actions to mean that one induces conjugate actions in each on subsets of the same measure, then this equivalence relation preserves entropy. Moreover, they showed that the conjugacy could be extended from the subsets to the rest of the two ambient spaces as an orbit equivalence precisely because the two sets have the same measure. An orbit equivalence arising in this fashion could be characterized in a variety of ways; in particular in ways that extended to higher dimensional actions. A variety of authors have pursued this area (Katok [27] for example). The last piece in this particular development is Hasfura-Buenaga's proof of an equivalence theorem in $\mathbb{Z}^d$ [16].

One could now see all three, orbit equivalence, even Kakutani equivalence, and conjugacy, as restricted orbit equivalences and look for the common thread in the corresponding equivalence theorems. This is what was attempted in [43]. The basic structure laid out there was that one should axiomatize a notion that measures how badly one is distorting an orbit. In both [43] and [25] this is based on an axiomatization of how one would measure the wildness of a permutation of a large block of the acting group $\mathbb{Z}$ or $\mathbb{Z}^d$. These axiomatizations were rather elaborate and technical, especially in [43].

Very quickly it became evident that the formulation in [43] was quite flawed. The author admits there that the basic structures will not lift reasonably to larger group actions, and natural examples arose of equivalence relations, that were restrictions on orbit equivalence, that "ought" to be but could not be brought into the framework of [43]. For example, α-equivalences described here in Examples 4 and 5 cannot arise from a 1-size. At a more subtle level, Fieldsteel indicated quite rightly that the definition given for an m-joining was not sufficiently robust. This is because an m-joining was required to be a joining perturbed by a "bounded coboundary", not by an m-equivalence. In particular an m-equivalence itself was not necessarily an m-joining.

The reason for this concern over basics is that as we indicated earlier there is a good deal more to the isomorphism theory than just Ornstein's theorem itself. There are all the constructive examples built to show it attains the best one could hope for. There is the "relativized theory" of

Thouvenot [55] leading to his deep study of the weak-Pinsker property [56]. In particular one has Thouvenot's result that the property of being Kakutani equivalent to a map with the weak-Pinsker property is inherited by factor actions. Fieldsteel and one of the authors generalized this to the [43] theory of entropy-preserving sizes [14]. One also has the theory of isometric and affine extensions of Bernoulli actions [23], [24], [46]. One could hope to generalize all of this work from conjugacy, and Bernoulli actions, to restricted orbit equivalences and m-finitely determined actions. This has been done for some examples [43]. Fieldsteel [12] showed that if one took compact group extensions of two ergodic actions, by the same group, so that the extensions were ergodic, then one could construct an orbit equivalence between the two extensions that preserved the group extension structures.

In a result which really deserves deeper study Fieldsteel and Friedman [13] showed that Belinskaya's amazing theorem (if the generating functions of an orbit equivalence were integrable, then the equivalence was essentially trivial) was false in higher dimensions even if *integrable* was replaced with *bounded*. This indicates that there is perhaps a non-trivial L^1 and even L^∞ orbit-equivalence theory in higher dimensions which is vacuous in 1. That is to say, there may be families of sizes in $\mathbb{Z}^d$ for all d, which for some d, say $d = 1$, give the same equivalence relations, but for larger d do not.

At this point the authors put forward the development in [25] as a restricted orbit equivalence theory for actions of $\mathbb{Z}^d$. This work is quite parallel to [43]. In [43] one takes an injection from a block of integers (i, j) and "pushes together" the range set to obtain a permutation of (i, j). In [25] one constructs permutations of boxes in $\mathbb{Z}^d$ from injections by "filling in" that is by taking those points that the map throws out of the box and placing them on the points in the box which have no preimage. These are similar, but far from the same notion. Certainly this new picture was much more generalizable, but still was tied to basic structures in the group $\mathbb{Z}^d$, in particular boxes.

In the meantime Ornstein and Weiss [37] had finished their seminal study of the ergodic theory of amenable group actions. This profound work makes it evident that the natural level at which the three basic tools of ergodic theory (Rokhlin lemma, ergodic theorem and a Shannon–McMillan theorem) apply is that of amenable groups. To be more precise, it is clear that they all apply at the level of discrete amenable groups. It still remains unclear to what degree the entropy theory for general amenable groups is well-founded. As our work on orbit equivalence will certainly also have technical problems at the level of uncountable groups, we gladly limited our work here to discrete amenable groups.

1.4 Directions for further study

We have indicated a number of areas that deserve further study in the light of our work here. We sketch a few of these here.

1. A relativized theory parallel to that of Thouvenot is probably directly accessible. We propose a natural notion for this. If m is a size, then in any given action $(X, \mathscr{F}, \mu, T)$, with H some fixed finite partition of X, consider the subgroup Γ_H of the full-group of elements with the property that for μ-a.e. x, both x and $\phi(x)$ lie in the same element of H. This is an L^1-closed subgroup. We say two arrangements are H relatively m-equivalent if they are m-equivalent via a sequence of elements ϕ_i in Γ_H.

2. The "non-example" described in the next section, based on work of Hoffman and Rudolph [20], proving an isomorphism theorem for certain endomorphisms is one direction for further work. Although these endomorphisms are measure-preserving, the natural orbit relation here is not, it is non-singular. Furthermore the natural group acting on these orbits is non-amenable, although the action is amenable. That is to say, to bring this example under a general umbrella such as we do here for the measure preserving case, one would have to consider a restricted orbit equivalence theory for non-singular amenable actions of perhaps non-amenable groups. This example and its generalizations (for example Kakutani equivalence for endomorphisms) makes this a significant direction for further study.

3. Following this direction further, so far we have always assumed that the orbits under consideration were organized as a copy of a group. One wants the orbits organized this way as our restrictions will all take the form of a metric on how badly this structure is distorted. One may consider other structures on a non-singular orbit relation. One idea is to replace the group with a lattice or graph structure. Some kind of stationarity of this structure would be needed. It could of course be a random structure, i.e. vary from point to point.

4. We have pointedly avoided non-discrete actions in this work but obviously one must address the technicalities in this for group actions, semigroup actions and perhaps some generalization to continuous graphs. A major issue in any such generalization is the exact nature of entropy. This can be avoided initially by considering entropy-zero classes. For

example, although the endomorphisms studied in [20] are of positive entropy all the entropy is carried by the trees of inverse images and relative to these trees the dynamics is zero entropy.

5. Certain notions exist that seem to have the character of either a restricted orbit equivalence, or of the finitely determined class in one. For example the weak-Pinsker property seems to have much the character of the finitely determined class of some entropy-preserving size. Is this the case? Feldman's notion of pre-Bernoulli also can perhaps be brought under this heading. Certainly one can look quite broadly for useful examples and applications of our general machinery.

2
Definitions and Examples

2.1 Orbits, arrangements and rearrangements

Let $(X, \mathscr{F}, \mu)$ be a fixed non-atomic Lebesgue probability space, that is to say it can be represented up to measure zero by Lebesgue measure on the unit interval. Let G be an infinite discrete amenable group. Suppose G acts ergodically on X as a group of measure preserving transformations. More precisely, we have measure-preserving maps T_g whose conjugation law is the group law of G ($T_{g_1} \circ T_{g_2} = T_{g_1 g_2}$) and for which all invariant sets have measure zero or one. Assume the action is free, that is to say for all $g \neq \text{id}$, $T_g(x) = x$ on at most a set of measure zero. (These are the standing hypotheses for the Ornstein–Weiss Rokhlin lemma.) Let $\mathcal{O} = \{x, T_g(x)\}_{g \in G} \subseteq X \times X$ be the orbit relation generated by this action. We say two such actions are **orbit-equivalent** if there is a measure-preserving map between their domains that carries one orbit relation to the other. We refer to such a map as an **orbit equivalence** between them. One case of the general theorem of Connes, Feldman and Weiss [6] is that all such orbit relations are orbit-equivalent. For our purposes then there is only one space and orbit relation. We study how these orbits can be arranged and rearranged to form orbits for an action.

Definition 2.1.1. *Let G be an infinite countable discrete amenable group. A* **G-arrangement** *α is any map from $\mathcal{O}$ to G that satisfies:*

(i) *α is 1-1 and onto, in that for a.e. $x \in X$, for all $g \in G$, there is a unique $x' \in X$ with $\alpha(x, x') = g$. We write $x' = T_g^{\alpha}(x)$;*

(ii) *α is measurable and measure preserving, i.e. for all $A \in \mathscr{F}, g \in G$, both $T_g^{\alpha}(A) \in \mathscr{F}$ and $\mu(T_g^{\alpha}(A)) = \mu(A)$; and*

(iii) *α satisfies the cocycle equation $\alpha(x_2, x_3)\alpha(x_1, x_2) = \alpha(x_1, x_3)$.*

As G will not vary for our considerations we will abbreviate this as an **arrangement**. *Let $\mathscr{A}$ denote the set of all such arrangements.*

Lemma 2.1.2. *A map α is a G-arrangement if and only if there is a measure preserving ergodic free action of G, T, whose orbit relation is $\mathcal{O}$ such that $\alpha(x, T_g(x)) = g$ for all $(x, T_g(x)) \in \mathcal{O}$.*

Thus the vocabulary of G-arrangements on $\mathcal{O}$ is precisely equivalent to the vocabulary of G-actions whose orbits are $\mathcal{O}$. For a G-arrangement α, we write T^α for the corresponding action. For a G-action T, we write α_T for the corresponding G-arrangement. This concept is very similar to that of a G-cocycle on $\mathcal{O}$. The cocycle would map the pair (x_1, g) to x_2 where the arrangment maps the pair (x_1, x_2) to g. Although the vocabulary of actions and cocycles is much more well known and all our work could be translated into either of these, the vocabulary of arrangements will make many formulations much simpler and more transparent.

Definition 2.1.3. *The* **full-group** *of $\mathcal{O}$ is the group (under composition) Γ of all measure-preserving invertible maps $\phi : X \to X$ such that for μ-a.e. $x \in X$, $(x, \phi(x)) \in \mathcal{O}$.*

Note that it would be sufficient in this definition to assume that ϕ is measurable and 1-1, since the fact that $\mathcal{O}$ is a measure-preserving orbit relation forces ϕ to be measure-preserving. Also note that the orbits of ϕ are a sub-relation of $\mathcal{O}$ and need not be all of $\mathcal{O}$.

Definition 2.1.4. *A* **G-rearrangement** *of $\mathcal{O}$ is a pair (α, ϕ), where α is a G-arrangement of $\mathcal{O}$ and $\phi \in \Gamma$. As G is fixed for our purposes we will abbreviate this to a* **rearrangement**. *Let $\mathscr{Q}$ denote the set of all such re-arrangements.*

Intuitively, a rearrangement is simply a change of an orbit from the arrangement α to the arrangement $\alpha\phi$, where $\alpha\phi(x, x') = \alpha(\phi(x), \phi(x'))$. One can formalize such a rearrangement in three different ways. Set $\mathscr{B}$ to be the set of bijections of G and $\mathscr{G}$ the subgroup of $\mathscr{B}$ fixing the identity. Both are topologized via the product topology on G^G. Notice there is a homomorphism $\hat{H} : \mathscr{B} \to \mathscr{G}$ given by $\hat{H}(q)(g) = q(g)q(\mathrm{id})^{-1}$. Observe that information is lost in mapping via $\hat{H}$. The kernel of $\hat{H}$ consists of the left translation maps.

To a rearrangment we can associate a family of functions $q_x^{\alpha,\phi} \in \mathscr{B}$ where

$$q_x^{\alpha,\phi}(g) = \alpha(x, \phi(T_g^\alpha(x))).$$

Now suppose α and β are two arrangements of the orbits $\mathcal{O}$. Regard the first as an initial and the second as a terminal arrangement. We can associate to this pair and any point x a bijection from G, fixing the identity that describes how the arrangement of the orbit has changed:

$$h_x^{\alpha,\beta}(g) = \beta(x, T_g^\alpha(x)).$$

Notice here that $\hat{H}(q_x^{\alpha,\phi}) = h_x^{\alpha,\alpha\phi}$.

Write $h^{\alpha,\beta} : X \to \mathscr{G}$.

The third way to view a rearrangement pair has a symbolic dynamic flavor. For each orbit $\mathcal{O}(x) = \{x'; (x, x') \in \mathcal{O}\}$, a rearrangement (α, ϕ) also gives rise to a natural map $G \to G$ (not a bijection though), given by

$$f_x^{\alpha,\phi}(g) = \alpha(T_g^\alpha(x)), \phi(T_g^\alpha(x)).$$

Visually, regarding $\mathcal{O}(x)$ laid out by α as a copy of G, ϕ translates the point at position g to position $f_x^{\alpha,\phi}(g)g$. Notice in particular that the definition of $f^{\alpha,\phi}$ is *stationary* in that

$$f_x^{\alpha,\phi}(g) = f_{T_g^\alpha(x)}^{\alpha,\phi}(\mathrm{id}).$$

Thus if we map a point x to the infinite word

$$w(x) = \{f_x^{\alpha,\phi}(g)\}_{g \in G} \in G^G$$

then

$$w(T_g(x)) = \sigma_g(w(x))$$

where σ_g is the **shift** action on G^G,

$$\sigma_g(h)(k) = h(kg).$$

There is a natural link between the three functions $h^{\alpha,\alpha\phi}$, $q^{\alpha,\phi}$ and $f^{\alpha,\phi}$ as follows. For any map $f : G \to G$ we define

$$Q(f)(g) = f(g)g \quad \text{and}$$

$$H(f)(g) = f(g)gf(\mathrm{id})^{-1}.$$

It is an easy calculation to see that

$$H(f^{\alpha,\phi}) = h^{\alpha,\alpha\phi}, \quad Q(f^{\alpha,\phi}) = q^{\alpha,\phi} \text{ and } H = \hat{H} \circ Q.$$

Let $\{F_i\}$ be a fixed Følner sequence for G. We will describe a number of concepts in terms of the F_i but will note at appropriate points where they are, in fact, independent of the Følner sequence.

Lemma 2.1.5. *Let $\{F_i\}$ be a Følner sequence in G, and let (α, ϕ) be a rearrangement. Define*

$$b_i(x) = \#\{g \in F_i; f_x^{\alpha,\phi}(g)g \notin F_i\} = \#\{g \notin F_i; f_x^{\alpha,\phi}(g)g \in F_i\}.$$

Then

$$\lim_{i\to\infty} \left\| \frac{b_i(x)}{\#F_i} \right\|_2 = 0.$$

Proof For a finite set $K \subseteq G$ set $E_K = \{x; f_x^{\alpha,\phi}(\mathrm{id}) \notin K\}$. For $\varepsilon > 0$ choose K so that $\mu(E_K) < \varepsilon/4$. Let $h_i(x) = \#\{g \in F_i; T_g^\alpha(x) \in E_K\}$. By the L^2-ergodic theorem, there exists I, such that for $i \geq I$,

$$\left\| \frac{h_i(x)}{\#F_i} \right\|_2 < \varepsilon/2.$$

By the Følner property, we may further select I such that for $i \geq I$,

$$\frac{\# \bigcup_{k\in K}(kF_i \triangle F_i)}{\#F_i} < \varepsilon/2.$$

Now for $i \geq I$, select $g \in F_i$ such that $f_x^{\alpha,\phi}(g)g \notin F_i$. For such g, either
(1) $f_x^{\alpha,\phi}(g) = f_{T_g^\alpha(x)}^{\alpha,\phi}(\mathrm{id}) \notin K$; or
(2) $f_x^{\alpha,\phi}(g)g = f_{T_g^\alpha(x)}^{\alpha,\phi}(\mathrm{id})g \in \bigcup_{k\in K}(kF_i \triangle F_i)$.

Thus $b_i(x) \leq h_i(x) + \# \bigcup_{k\in K}(kF_i \triangle F_i)$, so that $\limsup_{i\to\infty} \left\| \frac{b_i(x)}{\#F_i} \right\|_2 < \varepsilon$, which completes the proof. □

We now consider three metrics on the set of rearrangements. These all arise from natural topologies on functions $G \to G$, that is to say on G^G. As G is countable the only reasonable topology is the discrete one, using the discrete 0,1 valued metric. This topologizes G^G as a metrizable space with the product topology. This is the weakest topology for which the evaluations $g : f \to f(g)$ are continuous functions. Notice that H is a continuous map from G^G to itself and the map $h \to h^{-1}$ on $\mathcal{G}$ is continuous.

List the elements of G as $\{g_1 = \mathrm{id}, g_2, \dots\}$ and let d_0 be the 0,1 valued metric on G. Define a metric d on $\mathcal{G}$ as follows. Set

$$d(h_1, h_2) = \sum_i [d_0(h_1(g_i), h_2(g_i)) + d_0(h_1^{-1}(g_i), h_2^{-1}(g_i))]2^{-(i+1)}.$$

Notice that $d(h_1, h_2) \leq 2^{-i}$ if h_1, h_2, h_1^{-1}, and h_2^{-1} agree on $g_1, \ldots, g_i$. On the other hand if $d(h_1, h_2) < 2^{-i}$ then h_1, h_2 and their inverses agree on this list of i terms.

Lemma 2.1.6. *The metric d on $\mathscr{G}$ gives the restricted product topology and makes $\mathscr{G}$ a complete metric topological group.*

Proof It is evident that the group operations are continuous and that d gives the product topology. We show that d makes $\mathscr{G}$ complete. Suppose the h_i are a d-Cauchy sequence. It follows that for all $g \in G$, for all i sufficiently large, both $h_i(g)$ and $h_i^{-1}(g)$ remain constant. Hence $h_i \to h \in G^G$ and $h_i^{-1} \to k \in G^G$. But for any i sufficiently large

$$h_i(k(g)) = g$$

and then clearly $h \circ k = \mathrm{id}$ and $h \in \mathscr{G}$. □

A simple corollary of this is that $\mathscr{G}$ is a G_δ subset of G^G as we have just seen it to be topologically complete in the product topology.

We can use this to define an L^1 metric on arrangements:

$$\|\alpha, \beta\|_1 = \int d(h^{\alpha,\beta}, \mathrm{id})\, d\mu.$$

As $d(h_1 h_2, \mathrm{id}) \leq d(h_1, \mathrm{id}) + d(h_2, \mathrm{id})$ and $(h_x^{\alpha,\beta})^{-1} = h_x^{\beta,\alpha}$ we see that this is a metric.

Lemma 2.1.7. *The metric space $(\mathscr{A}, \|\cdot,\cdot\|_1)$ is complete.*

Proof Suppose the α_i form a Cauchy sequence of arrangements in $\|\cdot,\cdot\|_1$. There will then exist a subsequence $i(j)$ converging pointwise, i.e. so that for μ-a.e. $x \in X$, $h_x^{\alpha,\alpha_{i(j)}}$ are *Cauchy* in the metric d. As $(\mathscr{G}, d)$ is complete, we conclude that for μ-a.e. $x \in X$,

$$h_x^{\alpha,\alpha_{i(j)}} \underset{j}{\to} h_x.$$

Setting

$$\beta(x, T_g^\alpha(x)) = h_x(g)$$

it is straightforward to see that β is an arrangement. As $\|\alpha_{j(i)}, \beta\|_1 \to 0$ the proof is complete. □

We can also define a metric similar to d on G^G itself making it a complete metric space by just taking half of the terms in d:

$$d_1(f_1, f_2) = \sum_i d_0(f_1(g_i), f_2(g_i))2^{-i}.$$

This also leads to an L^1 metric on G^G-valued functions on a measure space:

$$\|f_1, f_2\|_1 = \int d_1(f_1, f_2)\, d\mu.$$

These two L^1 distances now give us two families of L^1 distances on the full-group, one a metric the other a pseudometric, associated with an arrangement α:

$$\begin{aligned}\|\phi_1, \phi_2\|_{\mathrm{w}}^{\alpha} &= \int d(h^{\alpha\phi_1,\alpha\phi_2}, \mathrm{id})\, d\mu = \|\alpha\phi_1, \alpha\phi_2\|_1 \\ &= \int d(H(f^{\alpha,\phi_1})^{-1} H(f^{\alpha,\phi_2}), \mathrm{id})\, d\mu,\end{aligned}$$

and

$$\begin{aligned}\|\phi_1, \phi_2\|_{\mathrm{s}}^{\alpha} &= \int d_1(f^{\alpha,\phi_1}, f^{\alpha,\phi_2})\, d\mu \\ &= \|f^{\alpha,\phi_1}, f^{\alpha,\phi_2}\|_1.\end{aligned}$$

The **weak** L^1 distance, $\|\cdot,\cdot\|_{\mathrm{w}}^{\alpha}$, is only a pseudometric but the **strong** L^1 distance, $\|\cdot,\cdot\|_{\mathrm{s}}^{\alpha}$, is a metric. Notice that T-invariance of μ tells us that

$$\begin{aligned}\|\phi_1, \phi_2\|_{\mathrm{s}}^{\alpha} &\leq 2 \int d_0(f^{\alpha,\phi_1}(\mathrm{id}), f^{\alpha,\phi_2}(\mathrm{id}))\, d\mu \\ &= 2\mu(\{x : \phi_1(x) \neq \phi_2(x)\}) \leq 2\|\phi_1, \phi_2\|_{\mathrm{s}}^{\alpha}.\end{aligned}$$

Thus, in fact, the topology generated by the strong L^1 distance on the full-group is independent of the arrangement α.

Before moving on to the weak* pseudometric we point out that it would perhaps be more correct to call the above weak and strong "distribution" metrics as they measure how closely the two rearrangements distribute the mass of X on the space of functions G^G. We of course call them L^1 pseudometrics as they arise from integrals of "distances".

To describe the weak*-distance between two arrangements we let $G^\star = G \cup \{\star\}$ be the one point compactification of G. Now $(G^\star)^G$ is a compact metric space and hence the Borel probability measures on $(G^\star)^G$, which we write as $\mathcal{M}_1(G^\star)$, are a compact and convex space in the weak*-topology (that is to say the topology induced on Borel measures as the dual of the continuous functions).

We can put an explicit metric on this space as follows. For any finite subset $F \subseteq G$ and $f \in G^G$ let f_F be the restriction of f to F. As f_F can be one of at most a countable collection of values, $f \to f_F$ partitions G^G into a countable collection of clopen sets. If two measures on $(G^\star)^G$ agree on these sets, that is to say on all cylinder sets that do not have a "$\star$" in their name, then they agree. Moreover the characteristic functions of these sets are continuous. Hence if $\mu_i(f_F) \to \mu(f_F)$ then $\mu_i \to \mu$ weak*. To turn this into a metric, let F_i be an increasing sequence of finite sets that exhaust G, for example a Følner sequence. For each F_i let $P(F_i)$ be the partition of G^G according to the values f_{F_i}. These partitions refine and for any fixed F, once $F \subseteq F_i$, f_F will be $P(F_i)$-measurable. Set

$$D(\mu_1, \mu_2) = \sum_i \Bigg(\sum_{p \in P(F_i)} |\mu_1(p) - \mu_2(p)| \Bigg) 2^{-(i+1)}.$$

Notice that

$$\sum_{p \in P(F_i)} |\mu_1(p) - \mu_2(p)|/2 \leq 1$$

and so the ith term in this sum is bounded by 2^{-i}. Moreover as the partitions $P(F_i)$ refine, the values

$$\sum_{p \in P(F_i)} |\mu_1(p) - \mu_2(p)|/2 \leq 1$$

increase. It follows that for all i

$$\sum_{p \in P(F_i)} |\mu_1(p) - \mu_2(p)| + 2^{-i} \geq D(\mu_1, \mu_2).$$

Before continuing we make two remarks. First it is clear from the discussion that this metric gives the weak*-topology on $(G^\star)^G$. Second, it is not too difficult to argue that those measures which put no support on $\star$ are a G_δ subset of the measures in $(G^\star)^G$. We will take a broader approach to this particular issue later in Section 7, showing that not only are the measures supported on G^G a G_δ but the weak*-topology here is independent of the way we choose to compactify G^G as long as the compactification is metric. At this point these issues are not important.

We now define the distribution pseudometric between rearrangements by

$$\|(\alpha, \phi), (\beta, \psi)\|_* = D((f^{\alpha,\phi})^*(\mu), (f^{\beta,\psi})^*(\nu)).$$

We can combine the two L^1-metrics on arrangements and the full-group to define a product metric on rearrangements in the form

$$\|(\alpha_1, \phi_1), (\alpha_2, \phi_2)\|_1 = \|\alpha_1, \alpha_2\|_1 + \mu(\{x : \phi_1(x) \neq \phi_2(x)\}).$$

We end this Section by relating this complete L^1-metric on rearrangements to the distribution pseudometric.

Lemma 2.1.8. *The map* $(\alpha, \phi) \to (f^{\alpha,\phi})^*(\mu)$ *is uniformly continuous as a map from* $(\mathcal{Q}, \|\cdot,\cdot\|_1)$ *to* $(\mathcal{Q}, \|\cdot,\cdot\|_*)$. *That is to say, given any* $\varepsilon > 0$ *there exists a* $\delta > 0$ *such that if* (α_1, ϕ_1) *and* (α_2, ϕ_2) *are two G-rearrangements that satisfy* $\|(\alpha_1, \phi_1), (\alpha_2, \phi_2)\|_1 < \delta$ *then*

$$\|(\alpha_1, \phi_2), (\alpha_2, \phi_2)\|_* < \varepsilon.$$

Proof Let $F \subseteq \{g_1, g_2, \ldots, g_K\}$ be any finite set. Suppose $\delta > 0$ and $\|\alpha_1, \alpha_2\|_1 < \delta/(K2^K)$. Then

$$\mu(\{x : h_x^{\alpha_1,\alpha_2}(g_i) \neq g_i \text{ for some } i \leq K\}) \leq \delta.$$

Thus,

$$\begin{aligned}
\mu(\{x; f_{x,F}^{\alpha_1,\phi_1} &\neq f_{x,F}^{\alpha_2,\phi_2}\}) \\
&\leq \#F\, \mu(\{x : \phi_1(x) \neq \phi_2(x)\}) \\
&\qquad + \mu(\{x : h_x^{\alpha_1,\alpha_2}(g_i) \neq g_i \text{ for some } i \leq K\}) \\
&\leq (\#F + K2^K)\|(\alpha_1, \phi_1), (\alpha_2, \phi_2)\|_1 \\
&< \delta(\#F + K2^K).
\end{aligned}$$

Thus for all i,

$$\|(\alpha_1, \phi_1), (\alpha_2, \phi_2)\|_* \leq (\#F_i + K_i 2^{K_i})\delta,$$

where $F_i \subseteq \{g_1, \ldots, g_{K_i}\}$.

Let $\varepsilon > 0$. Select i so that $2^{-i} < \varepsilon/2$ and let

$$\delta = \varepsilon/(2(\#F_i + 2^{K_i} K_i)).$$

The result follows. □

2.2 Definition of a size and m-equivalence

In this section we define the notion of a **size** m on rearrangements (α, ϕ) as a family of pseudometrics m_α on the full-group satisfying some simple relations to the metrics and pseudometrics defined in the previous Section. We then define the m-equivalence class of an arrangement

α to be those arrangements β for which the corresponding m_α and m_β-completions of the full-group are isometric in a canonical fashion.

We let Γ denote the full-group of $\mathcal{O}$, as before. Recall that $\mathcal{A}$ denotes the set of arrangements and $\mathcal{Q}$ denotes the set of rearrangements.

A size is a function

$$m : \mathcal{Q} \to \mathbb{R}^+$$

such that, if we write

$$m_\alpha(\phi_1, \phi_2) \underset{\text{defn}}{=} m(\alpha\phi_1, \phi_1^{-1}\phi_2),$$

then m satisfies the following three axioms.

Axiom 1. *For each $\alpha \in \mathcal{A}$, m_α is a pseudometric on Γ.*

Axiom 2. *For each $\alpha \in \mathcal{A}$, the identity map*

$$(\Gamma, m_\alpha) \xrightarrow{\text{id}} (\Gamma, \|\cdot,\cdot\|_{\text{w}}^{\alpha})$$

is uniformly continuous.

In particular if $m_\alpha(\phi_1, \phi_2) = 0$ then the two arrangements $\alpha\phi_1$ and $\alpha\phi_2$ are identical.

Axiom 3. *The function m is upper semi-continuous with respect to the distribution metric. That is to say, for every $\varepsilon > 0$, there exists a $\delta = \delta(\varepsilon, \alpha, \phi)$, such that if $\|(\alpha, \phi), (\beta, \psi)\|_* < \delta$ then $m(\beta, \psi) < m(\alpha, \phi) + \varepsilon$.*

This last condition tells us that if the two measures $(f^{\alpha,\phi})^*(\mu)$ and $(f^{\beta,\psi})^*(\nu)$ are the same, then $m(\alpha, \phi) = m(\beta, \psi)$. Hence the value m is well defined on those measures on G^G which arise as such an image, and we can write

$$m(\alpha, \phi) = m((f^{\alpha,\phi})^*(\mu)).$$

Later on we will find ourselves in the situation where the rearrangement (α, ϕ) is well-defined pointwise and the measure μ is allowed to vary. In this case we will be more specific and write $m_\mu(\alpha, \phi)$ or $m_{\alpha,\mu}(\phi_1, \phi_2)$.

Lemma 2.2.1. *Let m be a size. The identity map*

$$(\Gamma, \|\cdot,\cdot\|_{\text{s}}^{\alpha}) \xrightarrow{\text{id}} (\Gamma, m_\alpha)$$

is uniformly continuous.

Proof Let $\phi_1, \phi_2 \in \Gamma$. As

$$\|\phi_1, \phi_2\|_s^\alpha \geq \mu(\{x : \phi_1(x) \neq \phi_2(x)\})$$
$$= \|(\alpha\phi_1, \phi_1^{-1}\phi_2), (\alpha\phi_1, \mathrm{id})\|_1,$$

Lemma 2.1.8 tells us that for any $\delta_1 > 0$, there exists $\delta > 0$, such that if $\|\phi_1, \phi_2\|_s^\alpha < \delta$ then $\|(\alpha\phi_1, \phi_1^1\phi_2), \alpha\phi_1, \mathrm{id})\|_* < \delta_1$.

Fix an arrangement α_0 and $\phi_0 = \mathrm{id}$. Let $\varepsilon > 0$. Now select $\delta_1 = \delta(\varepsilon, \alpha_0, \mathrm{id})$ from Axiom 3. It follows that if $\|\phi_1, \phi_2\|_s^\alpha < \delta$ then

$$\|(\alpha\phi_1, \phi_1^1\phi_2), \alpha\phi_1, \mathrm{id})\|_* = \|(\alpha\phi_1, \phi_1^1\phi_2), \alpha_0, \mathrm{id})\|_* < \delta_1,$$

and hence

$$m_\alpha(\phi_1, \phi_2) = m(\alpha\phi_1, \phi_1^{-1}\phi_2) < \varepsilon.$$

□

Definition 2.2.2. *Let $\alpha \in \mathscr{A}$, $\{\phi_i\} \subseteq \Gamma$ and m be a size. Define Γ_α to be the equivalence classes of elements of Γ at an m_α distance zero. Thus $(\Gamma_\alpha, m_\alpha)$ is a metric space and we let $(\hat{\Gamma}_\alpha, m_\alpha)$ be its completion. That is to say, $\hat{\Gamma}_\alpha$ consists of sequences $\{\phi_i\}$ which are m_α-Cauchy, modulo the equivalence relation $\{\phi_i\} \sim \{\psi_i\}$ if*

$$m_\alpha(\phi_i, \psi_i) \underset{i}{\rightarrow} 0.$$

We write the elements of $\hat{\Gamma}_\alpha$ as $\langle\phi_i\rangle_\alpha$.

Lemma 2.2.3. *Fixing an arrangement α and size m the map $\phi \rightarrow \alpha\phi$ from Γ to $\mathscr{A}$ is well-defined as a map $\Gamma_\alpha \rightarrow \mathscr{A}$ and extends to a uniformly continuous map from $(\hat{\Gamma}_\alpha, m_\alpha) \rightarrow (\mathscr{A}, \|\cdot,\cdot\|_1)$. We refer to this map as $P_{m,\alpha}(\langle\phi_i\rangle_\alpha) = \lim_i \alpha\phi_i$.*

Proof As the map $\phi \rightarrow \alpha\phi$ is uniformly continuous $\Gamma_\alpha \rightarrow \mathscr{A}$ it extends to the completion of Γ_α (remember, $\mathscr{A}$ in the $\|\cdot,\cdot\|_1$-metric is complete).

□

Thus if $\langle\phi_i\rangle_\alpha \in \hat{\Gamma}_\alpha$ then $\alpha\phi_i \rightarrow \beta$ for some β in $\mathscr{A}$. Hence we will at times abbreviate this element $\langle\phi_i\rangle_\alpha \in \hat{\Gamma}_\alpha$ as $\hat{\beta}$, indicating that it is a lift to $\hat{\Gamma}_\alpha$ of the arrangement β.

The arrangements in the range of $P_{m,\alpha}$ are a first cut toward the m-equivalence class of α. The subset which is the actual equivalence class may be somewhat smaller.

Next, we see that since m satisfies Axiom 3, right multiplication in the full-group Γ is an isometry. Specifically, we have the following Lemma.

Lemma 2.2.4. *Let α be any arrangement. For all $\phi_0 \in \Gamma$, the map $\phi \to \phi\phi_0$ is an m_α-isometry, that is to say:*

$$m_\alpha(\phi_1\phi_0, \phi_2\phi_0) = m_\alpha(\phi_1, \phi_2).$$

Proof For all $\phi_0, \phi_1 \in \Gamma$, we have that

$$\|(\alpha\phi_0, \phi_0^{-1}\phi_1\phi_0), (\alpha, \phi_1)\|_* = 0.$$

Thus, by Axiom 3, for any $\phi_0, \phi_1, \phi_2 \in \Gamma$,

$$m(\alpha\phi_1\phi_0, \phi_0^{-1}\phi_1^{-1}\phi_2\phi_0) = m(\alpha\phi_1, \phi_1^{-1}\phi_2).$$

That is to say, $m_\alpha(\phi_1\phi_0, \phi_2\phi_0) = m_\alpha(\phi_1, \phi_2)$. □

It is not necessarily the case that left multiplication is continuous. In fact, left multiplication need not preserve the equivalence classes of the pseudometric m_α. For example, consider the size

$$m^0(\alpha, \phi) = \|\alpha, \alpha\phi\|_{\mathrm{w}}^{\alpha}.$$

For $G = \mathbb{Z}$ and $\phi_1 = T_1^\alpha$, we have that $m_\alpha^0(\mathrm{id}, \phi_1) = 0$, but $m_\alpha^0(\phi, \phi\phi_1) = m_\alpha^0(\mathrm{id}, \phi\phi_1\phi^{-1})$. Here one can prove that $m_\alpha^0(\mathrm{id}, \psi) = 0$ if and only if $\psi = T_k^\alpha$. So if $m_\alpha^0(\phi, \phi\phi_1) = 0$ then $\phi\phi_1\phi^{-1} = \phi T_1^\alpha \phi^{-1} = T_k^\alpha$. Thus, (using ergodicity and freeness), $k = 1$ and ϕ commutes with T_1^α.

Hence ϕ is itself T_n^α for some n, yet the full group is much larger than this.

We do have something akin to semi-continuity of right multiplication. Suppose $\|\alpha\phi_i, \beta\|_{\mathrm{w}}^{\alpha} \underset{i}{\to} 0$. Then, for a.e. x, we have that

$$\alpha(\phi_i\phi_1(x), \phi_i\phi_2(x)) \underset{i}{\to} \beta(\phi_1(x), \phi_2(x)),$$

and so $(\alpha\phi_i\phi_1, (\phi_i\phi_1)^{-1}(\phi_i\phi_2)) = (\alpha\phi_i\phi_1, \phi_1^{-1}\phi_2)$ converges in distribution to $(\beta\phi_1, \phi_1^{-1}\phi_2)$. Axiom 3 now tells us that

$$\lim_{i\to\infty} m_\alpha(\phi_i\phi_1, \phi_i\phi_2) \le m_\beta(\phi_1, \phi_2). \qquad (\#)$$

Definition 2.2.5. *We say that m is a* **3^+ size** *if whenever $\{\phi_i\}$ is m_α-Cauchy and hence $\|\alpha\phi_i, \beta\|_1 \underset{i}{\to} 0$, the inequality (#) is an equality.*

In particular, this will be true if Axiom 3 is replaced by the following:

Axiom 3$^+$. *Let α be any ordering, $\phi \in \Gamma$ and $\varepsilon > 0$. There exists δ so that if $\|(\alpha,\phi),(\beta,\psi)\|_* < \delta$, then $|m(\alpha,\phi) - m(\beta,\psi)| < \varepsilon$; i.e. m is continuous in the distribution metric.*

The isometry of right multiplication on Γ (Lemma 2.2.4), implies that the action of Γ by right multiplication extends as an isometric action of Γ on $(\hat{\Gamma}_\alpha, m_\alpha)$.

This action is clearly topologically transitive, since $\hat{\Gamma}$ is an orbit closure. In particular, we get the following:

Lemma 2.2.6. *Let α be an arrangement and m a size. For any $\langle\phi_i\rangle_\alpha$ in $(\hat{\Gamma}_\alpha, m_\alpha)$,*

$$\lim_{j\to\infty} \hat{m}_\alpha(\langle\langle\phi_i\rangle_\alpha \phi_j^{-1}\rangle_\alpha, \langle \mathrm{id}\rangle_\alpha) = 0.$$

Proof Let $\{\phi_i\}$ be an m_α-Cauchy sequence, (representing the class $\langle\phi_i\rangle_\alpha$). Fix j and using Lemma 2.2.4 compute that

$$\begin{aligned} m_\alpha(\langle\langle\phi_i\rangle_\alpha \phi_j^{-1}\rangle_\alpha, \langle \mathrm{id}\rangle_\alpha) &= m_\alpha(\langle\phi_i \phi_j^{-1}\rangle_\alpha, \langle \mathrm{id}\rangle_\alpha) \\ &= \lim_{i\to\infty} m_\alpha(\phi_i\phi_j^{-1}, \mathrm{id}) \\ &= \lim_{i\to\infty} m_\alpha(\phi_i, \phi_j). \end{aligned}$$

Since $\{\phi_i\}$ is m_α-Cauchy, for any $\varepsilon > 0$, there exists I such that for all $i, j \geq I$, $m_\alpha(\phi_i, \phi_j) < \varepsilon$. Thus, for all $j \geq I$,

$$\lim_{i\to\infty} m_\alpha(\phi_i, \phi_j) < \varepsilon.$$

The result follows. □

In particular, this tells us that the action of the full-group Γ on $(\hat{\Gamma}_\alpha, m_\alpha)$ is minimal, i.e. every orbit is dense.

For a size m, we would like to define a notion of m-equivalence between two arrangements α and β. Within the context of rearrangements, a natural candidate would be to say that two arrangements α and β are m-equivalent if β is in the range of $P_{m,\alpha}$. That is, there is a sequence of rearrangements (α, ϕ_i) with $\alpha\phi_i \to \beta$ in L^1, such that $\{\phi_i\}$ is m_α-Cauchy. The problem with this definition is that on the face of it the "equivalence relation" is not symmetric. More precisely, it is not clear that, in general, the m_α-Cauchiness of $\{\phi_i\}$ will imply m_β-Cauchiness of $\{\phi_i^{-1}\}$.

The following theorem describes this situation more precisely.

Theorem 2.2.7. *Suppose α is an arrangement, m is a size, $\{\phi_i\}$ is m_α-Cauchy and $\alpha\phi_i \underset{i}{\rightarrow} \beta$ in L^1. The map*

$$P : (\Gamma, m_\beta) \rightarrow (\hat{\Gamma}_\alpha, m_\alpha),$$

given by

$$P(\phi) = \langle \phi_i \phi \rangle_\alpha$$

is a Γ-equivariant contraction, so that

$$m_\alpha(P(\phi_1), P(\phi_2)) \leq m_\beta(\phi_1, \phi_2).$$

Hence P extends to a Γ-equivariant contraction

$$P : (\hat{\Gamma}_\beta, m_\beta) \rightarrow (\hat{\Gamma}_\alpha, m_\alpha).$$

The following are equivalent:

(1) id $\in$ Range(P);
(2) *P is onto;*
(3) *P is an isometry;*
(4) *$\{\phi_i^{-1}\}$ is m_β-Cauchy.*

Lastly, if m is a 3^+ size, then for all β in the range of $P_{m,\alpha}$, P is an isometry.

Proof That P is a Γ-equivariant contraction we verified earlier (in (#)) as a consequence of Axiom 3. This certainly implies that P extends to a Γ-equivariant contraction

$$P : (\hat{\Gamma}_\beta, m_\beta) \rightarrow (\hat{\Gamma}_\alpha, m_\alpha).$$

Moving on to the four equivalent statements: if id $\in$ Range(P) then there exists $\hat{\gamma} \in (\hat{\Gamma}_\beta, m_\beta)$ such that $P(\hat{\gamma}) = \text{id}$. Suppose $\hat{\gamma} = \langle \psi_i \rangle_\beta$. To show that P is onto, we need only show that P maps onto Γ.

Let $\phi \in \Gamma$. We will show that $P(\hat{\gamma}\phi) = \phi$.

In fact, $P(\hat{\gamma}\phi) = \langle P(\psi_i \phi)_i \rangle_\alpha$. We need only prove that

$$\lim_{i \to \infty} P(\psi_i \phi) = \phi.$$

Compute that

$$\begin{aligned}
\lim_{i \to \infty} m_\alpha(P(\psi_i \phi), \phi) &= \lim_{i \to \infty} m_\alpha(\langle \{\phi_j \psi_i \phi\}_j \rangle_\alpha, \phi) \\
&= \lim_{i \to \infty} \lim_{j \to \infty} m_\alpha(\phi_j \psi_i \phi, \phi) \\
&= \lim_{i \to \infty} \lim_{j \to \infty} m_\alpha(\phi_j \psi_i, \text{id})
\end{aligned}$$

since Γ acts isometrically,

$$
\begin{aligned}
&= \lim_{i\to\infty} m_\alpha(\langle\{\phi_j\psi_i\}_j\rangle_\alpha, \mathrm{id})\\
&= \lim_{i\to\infty} m_\alpha(\langle\{\phi_j\}\rangle_\alpha\psi_i, \mathrm{id})\\
&= \lim_{i\to\infty} m_\alpha(P(\psi_i), \mathrm{id})\\
&= 0,
\end{aligned}
$$

since $P(\hat{\gamma}) = \mathrm{id}$.

Hence if $\mathrm{id} \in \mathrm{Range}(P)$ then $\Gamma \subseteq \mathrm{Range}(P)$ and P must be onto. Thus (1) and (2) are equivalent.

Next, we argue that if $\mathrm{id} \in \mathrm{Range}(P)$ then the sequence $\{\phi_i^{-1}\}$ must be m_β-Cauchy.

Suppose $P(\hat{\gamma}) = \mathrm{id}$, where $\hat{\gamma} = \langle\psi_i\rangle_\beta$. The fact that P is an equivariant contraction implies that

$$\langle\phi_i\rangle_\alpha = \langle\psi_i^{-1}\rangle_\alpha.$$

To see this, simply compute that

$$
\begin{aligned}
\lim_{i\to\infty} m_\alpha(\phi_i, \psi_i^{-1}) &= \lim_{i\to\infty} m_\alpha(\phi_i\psi_i^{-1}, \mathrm{id})\\
&= \lim_{i\to\infty} m_\alpha(\langle\phi_j\rangle_\alpha\psi_i, \mathrm{id})\\
&= \lim_{i\to\infty} m_\alpha(P(\psi_i), P(\hat{\psi}))\\
&\le \lim_{i\to\infty} m_\beta(\psi_i, \hat{\gamma}),\\
&= 0.
\end{aligned}
$$

Define

$$Q : (\Gamma, m_\alpha) \to (\hat{\Gamma}_\beta, m_\beta),$$

by

$$Q(\phi) = \langle\psi_i\phi\rangle_\beta.$$

Exactly as for P, argue that Q is a Γ-equivariant contraction, so that

$$\hat{m}_\beta(Q(a), Q(b)) \le m_\alpha(a, b).$$

Hence Q extends as a Γ-equivariant contraction

$$Q : (\hat{\Gamma}, m_\alpha) \to (\hat{\Gamma}, m_\beta).$$

Since $\hat{\gamma} \in (\hat{\Gamma}, m_\beta)$, by Lemma 2.2.4, we see that

$$Q(\langle \psi_i^{-1} \rangle_\alpha) = \text{id}.$$

The above argument (applied to Q) now shows that

$$\langle \phi_i^{-1} \rangle_\beta = \langle \psi_i \rangle_\beta.$$

In particular this implies that $\{\phi_i^{-1}\}$ is m_β-Cauchy. Thus, since Q is a contraction, for $\phi \in \Gamma$,

$$m_\beta(Q(P(\phi)), \phi) = 0.$$

Hence for all $\hat{\phi} \in (\hat{\Gamma}_\beta, \hat{m}_\beta)$, we see that

$$m_\beta(Q(P(\hat{\phi})), \hat{\phi}) = 0.$$

Thus, for $a, b \in (\hat{\Gamma}_\beta, \hat{m}_\beta)$,

$$\begin{aligned} m_\beta(a,b) &= m_\beta(Q(P(a)), Q(P(b))) \\ &\leq m_\alpha(P(a), P(b)) \\ &\leq m_\beta(a,b), \end{aligned}$$

so that P is an isometry.

If P is an isometry, then of course, $\{\phi_i^{-1}\}$ is m_β-Cauchy, by Lemma 2.2.4. Hence, id $\in$ Range(P). This completes the proof that statements (1)–(4) are equivalent.

If m is a 3^+ size, then P is directly seen to be an isometry for all β in the range of $P_{m,\alpha}$. □

Definition 2.2.8. *There are three natural levels now on which to define* **m-equivalence classes**. *The first is the most functorial, as a subset of* $\hat{\Gamma}_\alpha$ *we can set*

$$\hat{E}_m(\alpha) = \{\langle \phi_i \rangle_\alpha \in \hat{\Gamma}_\alpha : \alpha\phi_i \to \beta \text{ and } \langle \phi_i^{-1} \rangle_\beta \in \hat{\Gamma}_\beta\}.$$

Second, we can consider the relation on arrangements given by

$$E_m(\alpha) = \{\beta | \alpha \overset{m}{\sim} \beta\} = P_{m,\alpha}(\hat{E}_m(\alpha)).$$

Third, we can consider the category of free and ergodic G-actions T and S and say S is **m-equivalent** *to T if there is an arrangement $\beta \in E_m(\alpha_T)$ with T^β conjugate to S. We indicate all three of these relations by the symbol $\overset{m}{\sim}$. Thus we will write $\langle \phi_i \rangle_\alpha \overset{m}{\sim} \langle \psi_i \rangle_\alpha$, $\alpha \overset{m}{\sim} \beta$ and $T \overset{m}{\sim} S$.*

We investigate the first two of these. We will show that $\hat{E}_m(\alpha)$ is a dense G_δ subset of $\hat{\Gamma}_\alpha$ directly by exhibiting it as a countable intersection of open sets. We will also show that each equivalence class $E_m(\alpha)$, as a subset of the arrangements, can be endowed with a natural m_α-metric making it a universal G_δ. We obtain this latter result by showing the map $\hat{E}_m(\alpha) \to E_m(\alpha)$ is obtained by considering $\hat{E}_m(\alpha)$ modulo a natural group of isometries of $\hat{E}_m(\alpha)$.

Our first task is to see that on the level of arrangements $\overset{m}{\sim}$ is an equivalence relation. It then follows automatically for free and ergodic actions.

We begin by putting a natural metric on $E_m(\alpha)$. As $P_{m,\alpha}$ is continuous from $\hat{\Gamma}_\alpha \to \mathscr{A}$, for any β in the range of $P_{m,\alpha}$, its pull-back $P^{-1}_{m,\alpha}(\beta)$ will be a closed set. We can use the Hausdorf metric on these closed sets to put a metric on the equivalence class $E_m(\alpha)$. More precisely, for $\beta_1, \beta_2 \in E_m(\alpha)$, define

$$m(\beta_1, \beta_2) = \inf_{\substack{\langle \phi_i \rangle_\alpha \in P^{-1}_{m,\alpha}(\beta_1) \\ \langle \psi_i \rangle_\alpha \in P^{-1}_{m,\alpha}(\beta_2)}} \hat{m}_\alpha(\langle \phi_i \rangle_\alpha, \langle \psi_i \rangle_\alpha) = m_\alpha(P^{-1}_{m,\alpha}(\beta_1), P^{-1}_{m,\alpha}(\beta_2)).$$

At this point it is not quite evident that this is a metric and not just a pseudometric. We have the following trivial consequence to Axiom 2.

Lemma 2.2.9. *For every $\varepsilon > 0$, there exists a $\delta > 0$, such that if $m(\alpha, \beta) < \delta$ then $\|\alpha, \beta\|_1 < \varepsilon$. That is to say, the identity is uniformly continuous from m to $\|\cdot, \cdot\|_1$. Hence $m(\alpha, \beta)$ is a metric.*

In particular, $\alpha \overset{m}{\sim} \beta$ if the map P, defined in Theorem 2.2.7, is an isometry.

We can now put this together in a simple form:

Lemma 2.2.10. *There exists a Γ-equivariant isometry $P : (\hat{\Gamma}, \hat{m}_\alpha) \to (\hat{\Gamma}, \hat{m}_\beta)$, with $P_{m,\beta} P = P_{m,\alpha}$, if and only if $\alpha \overset{m}{\sim} \beta$.*

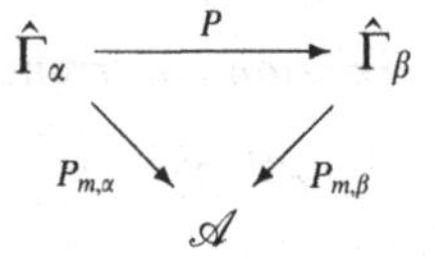

Proof The existence of such a P follows from Theorem 2.2.7 if $\alpha \overset{m}{\sim} \beta$. Conversely, suppose such a P exists. As $P_{m,\alpha}(\text{id}) = \alpha$, we see that $P_{m,\beta}(P(\text{id})) = \alpha$. Now $P(\text{id}) = \langle \phi_i^{-1} \rangle_\beta \in (\hat{\Gamma}_\beta, m_\beta)$. Thus, $P_{m,\beta}(\langle \phi_i^{-1} \rangle_\beta) = \alpha$.

For any j, we compute that

$$\begin{aligned} m_\alpha(P^{-1}(\mathrm{id}), \phi_j) &= m_\beta(P(P^{-1}(\mathrm{id})), P(\phi_j)) \\ &= m_\beta(\mathrm{id}, \langle \phi_i^{-1} \rangle_\beta \phi_j) \\ &\to 0 \text{ in } j, \text{ by Lemma 2.2.4.} \end{aligned}$$

Thus $P^{-1}(\mathrm{id}) = \langle \phi_j \rangle_\alpha \in (\hat{\Gamma}_\alpha, m_\alpha)$, and id $\in$ Range(P). □

The following theorem is now evident.

Theorem 2.2.11. *The relation $\overset{m}{\sim}$ on $\mathscr{A}$ is an equivalence relation.*

From this we see that $\overset{m}{\sim}$ breaks $\mathscr{A}$ into disjoint equivalence classes on each of which we have defined a metric. Our final step is to see that relative to this metric each of these classes is a Polish space. We begin with the classes $\hat{E}_m(\alpha)$.

Theorem 2.2.12. *The set $\hat{E}_m(\alpha)$ is a G_δ-subset of $\hat{\Gamma}_\alpha$. As the full-group is m_α-separable the m_α-topology on $\hat{E}_m(\alpha)$ is Polish.*

Proof For any ϕ and ψ in the full-group and $\varepsilon > 0$ let

$$\mathcal{O}(\phi, \psi, \varepsilon) = \{\langle \phi_i \rangle_\alpha \in \hat{\Gamma}_\alpha : \alpha\phi_i \to \beta \text{ and } m_\beta(\phi, \psi) < m_\alpha(\langle \phi_i \rangle_\alpha \phi, \langle \phi_i \rangle_\alpha \psi) + \varepsilon\}.$$

Notice that:

(1) $\hat{E}_m(\alpha) \subseteq \mathcal{O}(\phi, \psi, \varepsilon)$ for all ϕ, ψ and ε; and
(2) if $\langle \phi_i \rangle_\alpha \in \mathcal{O}(\phi, \psi, \varepsilon)$ for all ϕ, ψ and ε then

$$m_\beta(\phi, \psi) \le m_\alpha(P(\phi), P(\psi)).$$

As we always have $m_\beta(\phi, \psi) \ge m_\alpha(P(\phi), P(\psi))$, we conclude P is an isometry and hence $\langle \phi_i \rangle_\alpha \in \hat{E}_\alpha$.

The full-group is separable in the L^1-topology and so, by Axiom 2, is separable in the m_α-topology and we can find a countable collection $\{\phi_i\}$ dense in all the m_α or m_β-topologies. Hence

$$\cap_{i,j,k} \mathcal{O}(\phi_i, \phi_j, 1/k) = \hat{E}_m(\alpha).$$

It remains to see that the sets $\mathcal{O}(\phi, \psi, \varepsilon)$ are open in $\hat{\Gamma}_\alpha$. Suppose $\langle \phi_i \rangle_\alpha \in \mathcal{O}(\phi, \psi, \varepsilon)$ and hence there is an $e > 0$ with

$$m_\beta(\phi, \psi) < m_\alpha(\langle \phi_i \rangle_\alpha \phi, \langle \phi_i \rangle_\alpha \psi) + \varepsilon + e.$$

By Axiom 2, Lemma 2.1.8 and Axiom 3 there is a $\delta_0 > 0$ such that if

$$m_\alpha(\langle\phi_i\rangle_\alpha, \langle\phi_i'\rangle_\alpha) < \delta_0$$

then $\|\beta, \beta'\|_1$ will be sufficiently small to imply that

$$m_{\beta'}(\phi, \psi) < m_\beta(\phi, \psi) - e/3.$$

As

$$\begin{aligned} m_\alpha(\langle\phi_i\rangle_\alpha\phi, \langle\phi_i\rangle_\alpha\psi) & \\ &\leq m_\alpha(\langle\phi_i'\rangle_\alpha\phi, \langle\phi_i'\rangle\psi) + m_\alpha(\langle\phi_i\rangle_\alpha, \langle\phi_i'\rangle_\alpha\phi) \\ &\qquad + m_\alpha(\langle\phi_i\rangle_\alpha\psi, \langle\phi_i'\rangle_\alpha\psi) \\ &= m_\alpha(\langle\phi_i'\rangle_\alpha\phi, \langle\phi_i'\rangle_\alpha\psi) + 2m_\alpha(\langle\phi_i\rangle_\alpha, \langle\phi_i'\rangle_\alpha), \end{aligned}$$

making sure that $\delta_0 < e/3$ we will have

$$\begin{aligned} m_{\beta'}(\phi, \psi) &< m_\beta(\phi, \psi) - e/3 \\ &< m_\alpha(\langle\phi_i\rangle_\alpha\phi, \langle\phi_i\rangle_\alpha\psi) + \varepsilon + 2e/3 \\ &< m_\alpha(\langle\phi_i'\rangle_\alpha\phi, \langle\phi_i'\rangle_\alpha\psi) + \varepsilon \end{aligned}$$

and $\beta' \in \mathcal{O}(\phi, \psi, \varepsilon)$. □

Suppose that $\hat{\beta}_1 = \langle\phi_i^1\rangle_\alpha$ and $\hat{\beta}_2 = \langle\phi_i^2\rangle_\alpha$ are in $\hat{E}_m(\alpha)$ with $\alpha\phi_i^1 \to \beta$ and $\alpha\phi_i^2 \to \beta$. That is to say, $P_{m,\alpha}(\hat{\beta}_1) = P_{m,\alpha}(\hat{\beta}_2) = \beta$. This means that for all ϕ and ψ that

$$m_\alpha(\hat{\beta}_1\phi, \hat{\beta}_1\psi) = m_\beta(\phi, \psi) = m_\alpha(\hat{\beta}_2\phi, \hat{\beta}_2\psi).$$

Begin the definition of an m_α-isometry $I_{\hat{\beta}_1, \hat{\beta}_2}$ by setting

$$I_{\hat{\beta}_1, \hat{\beta}_2}(\hat{\beta}_1\phi) = \hat{\beta}_2\phi.$$

The above calculation implies that $I_{\hat{\beta}_1, \hat{\beta}_2}$ is an m_α-isometry where it is defined, and, as the $\hat{\beta}_1\phi$ are dense in $\hat{\Gamma}_\alpha$, $I_{\hat{\beta}_1, \hat{\beta}_2}$, will extend to an isometry of $\hat{\Gamma}_\alpha$. Notice that this makes $I_{\hat{\beta}_1, \hat{\beta}_2}$ commute with the action of e full group on $\hat{\Gamma}_\alpha$.

Lemma 2.2.13. *For all $\hat{\beta}_1$, $\hat{\beta}_2$ in $\hat{E}_m(\alpha)$ with $P_{m,\alpha}(\hat{\beta}_1) = P_{m,\alpha}(\hat{\beta}_2)$, we have*

$$P_{m,\alpha} I_{\hat{\beta}_1, \hat{\beta}_2} = P_{m,\alpha}.$$

Proof As $P_{m,\alpha}$ is equivariant with the action of the full-group,

$$P_{m,\alpha} I_{\hat{\beta}_1, \hat{\beta}_2}(\hat{\beta}_1\phi) = P_{m,\alpha}(I_{\hat{\beta}_1, \hat{\beta}_2}(\hat{\beta}_1\phi)) = \beta\phi = P_{m,\alpha}(\hat{\beta}_1\phi).$$

This now extends to all of $E_m(\alpha)$ as the $\hat{\beta}_1\phi$ are dense. □

Definition 2.2.14. *Let $\mathscr{I}$ consist of those m_α-isometries of $\hat{\Gamma}_\alpha$ commuting with the action of the full-group and satisfying*

$$P_{m,\alpha} I = P_{m,\alpha}.$$

This is a complete and separable metrizable space under pointwise convergence. Notice then that for any $\hat{\beta}_1 \in \hat{E}_m(\alpha)$ and $I \in \mathscr{I}$, setting $\hat{\beta}_2 = I(\hat{\beta}_1)$, we will have $\hat{\beta}_2 \in \hat{E}_m(\alpha)$ and

$$I = I_{\hat{\beta}_1, \hat{\beta}_2}.$$

Furthermore, as right multiplication by elements of the full-group is an m_α-isometry of $\hat{\Gamma}_\alpha$,

$$m_\alpha(\hat{\beta}, I(\hat{\beta})) = C(\hat{\beta})$$

is a constant on $\hat{\Gamma}_\alpha$.

We now argue that the orbits of $\mathscr{I}$ are closed sets. Suppose $I_i(\hat{\beta}_0)$ converges to some $\hat{\beta}_1$. Then in particular the $I_i(\hat{\beta}_0)$ will be m_α-Cauchy. But the above remark implies that $I_i(\hat{\beta})$ is m_α-Cauchy for all $\hat{\beta}$ and in particular converges to some $I(\hat{\beta}) \in \hat{\Gamma}_\alpha$. That is to say, $I_i \to I$ uniformly and we conclude that $I \in \mathscr{I}$ and all orbits are closed.

Lemma 2.2.15. *Using the infimum metric, the space $\hat{\Gamma}_\alpha / \mathscr{I}$ is a complete separable metric space.*

Proof To see that the infimum is a metric suppose

$$\inf_{I_1, I_2 \in \mathscr{I}} (m_\alpha(I_1(\hat{\beta}_1), I_2(\hat{\beta}_2))) = 0$$

then of course

$$\inf_{I \in \mathscr{I}} (m_\alpha(\hat{\beta}_1, I(\hat{\beta}_2))) = 0$$

and there will be a sequence $I_i(\hat{\beta}_1) \to \hat{\beta}_2$. But this says $I_i(\hat{\beta}_1)$ is m_α-Cauchy, and hence $I_i(\hat{\beta})$ is m_α-Cauchy for all $\hat{\beta}$. That is to say, the I_i converge uniformly to another $I \in \mathscr{I}$ implying $\hat{\beta}_2 = I(\hat{\beta}_1)$.

Thus the infimum is a metric on $\hat{\Gamma}_\alpha / \mathscr{I}$.

Separability follows from the fact that $\hat{\Gamma}_\alpha$ is separable. □

We now show that $E_m(\alpha)$ is also a G_δ-subset of $\mathscr{A}$ by showing that it embeds as $\hat{E}_m(\alpha)/\mathscr{I}$ which we show to be a residual subset of $\hat{\Gamma}_\alpha / \mathscr{I}$. We achieve the embedding by lifting $\beta \in E_m(\alpha)$ to $P_{m,\alpha}^{-1}(\beta) \cap \hat{E}_m(\alpha)$ which we note maps to a singleton in $\hat{\Gamma}_\alpha / \mathscr{I}$. It follows directly from the definitions that this is a continuous embedding. This implies $E_m(\alpha)$ is a universal G_δ, that is to say a G_δ-subset of any metric space in which it is embedded.

Lemma 2.2.16. *The sets $\mathcal{O}(\phi, \psi, \varepsilon)$ of Theorem 2.2.12 are $\mathcal{I}$-invariant. Hence*

$$\hat{E}_m(\alpha)/\mathcal{I} = \cap_{i,j,k}\mathcal{O}(\phi_i, \phi_j, 1/k)/\mathcal{I}$$

is a G_δ-subset.

Proof Remember that

$$\mathcal{O}(\phi, \psi, \varepsilon) = \{\hat{\beta} : P_{m,\alpha}(\hat{\beta}) = \beta \text{ and } m_\beta(\phi, \psi) < m_\alpha(\hat{\beta}\phi, \hat{\beta}\psi) + \varepsilon\}.$$

For $I \in \mathcal{I}$ we have both $P_{m,\alpha}I = P_{m,\alpha}$ and I is an m_α-isometry. These combine to say that if $\hat{\beta}' = I(\hat{\beta})$ then $P_{m,\alpha}(\hat{\beta}') = \hat{\beta}$ and

$$m_\alpha(\hat{\beta}'\phi, \hat{\beta}'\psi) = m_\alpha(I(\hat{\beta}\phi), I(\hat{\beta}\psi)) = m_\alpha(\hat{\beta}\phi, \hat{\beta}\psi).$$

It is now clear that

$$I(\mathcal{O}(\phi, \psi, \varepsilon)) = \mathcal{O}(\phi, \psi, \varepsilon).$$

It is obvious that if $\mathcal{O}$ is an $\mathcal{I}$-invariant open set then $\mathcal{O}/\mathcal{I}$ is open, and that the collection of $\mathcal{I}$-invariant sets form a Boolean algebra, with moding out by $\mathcal{I}$-equivariant with the Boolean operations. This completes the result. □

Perhaps the best heuristic to take away from this Section is the image of the m-equivalence classes as foliating the set of arrangements. Each leaf of this foliation is metrized by its m_α as a Polish space and Γ acts on each leaf minimally and isometrically.

2.3 Seven examples

Having developed the axiomatics of m-equivalence we now give a list of examples to indicate the range of equivalence relations that can be brought under this perspective. In [43] a number of examples and classes of examples are discussed. Some of those are quite speculative. The Appendix of this work demonstrates how to bring all those examples under the umbrella we open here. The examples we discuss in this Section are those which are most obviously significant and directly related to classical issues in ergodic theory. As part of this discussion we give some general principles that underlie many of these examples as the beginning of we expect a fruitful study of what a size might look like in general.

Many examples of sizes have the common feature of being integrals of some pointwise calculation of the distortion of a single orbit. To make this precise we first review some material about bijections of G. Remember that $\mathscr{B}$ is the space of all bijections of the group G with the product topology, $\mathscr{G}$ is the space of bijections fixing id and we metrized both with a complete metric d. The group G can be regarded as a subgroup of $\mathscr{B}$ acting by left multiplication, $(g(g') = gg')$. The map $\hat{H} : \mathscr{B} \to \mathscr{G}$ given by $\hat{H}(q) = qq(\mathrm{id})^{-1}$ is a contraction in d. Also G acting by right multiplication conjugates $\mathscr{B}$ to itself giving an action of G on $\mathscr{B}$. $(T_g(q)(g') = q(g'g)g^{-1}.)$ We view this action by representing an element $q \in \mathscr{B}$ by a map $f : G \to G$, $f(g) = q(g)g^{-1}$. Those maps $f \in G^G$ that arise from bijections are a G_δ and hence a Polish space we call F. The map $q \to f$ is obviously a homeomorphism from $\mathscr{B}$ to F. For $f \in F$ let $Q(f)$ be the associated bijection and for $q \in \mathscr{B}$ let $F(q)$ be the associated name in G^G. The action of G on $\mathscr{B}$ in its representation as F is the shift action $\sigma_g(f)(g') = f(g'g)$. Any rearrangement pair (α, ϕ) then gives rise to an ergodic shift invariant measure on this Polish subset of G^G and any ergodic shift invariant measure is an ergodic action of G with a canonical rearrangement pair. The probability measures on a Polish space are weak* Polish [57] and hence the invariant and ergodic measures on this Polish space are weak* Polish.

We will now define a general class of sizes that arise as integrals of valuations made on the bijections $q_x^{\alpha,\phi}$.

Definition 2.3.1. *A Borel* $D : \mathscr{B} \to \mathbb{R}^+$ *is called a* **size kernel** *if it satisfies:*

(1) $D(q) \geq 0$;
(2) $D(\mathrm{id}) = 0$;
(3) $D(q(\mathrm{id})^{-1}q^{-1}q(\mathrm{id})) = D(q)$;
(4) $D(q_1(\mathrm{id})q_2q_1^{-1}(\mathrm{id})q_1) \leq D(q_1) + D(q_2)$;
(5) *for every* $\varepsilon > 0$ *there is a* $\delta > 0$ *so that if* $D(q) < \delta$ *then* $d(\mathrm{id}, H(q)) < \varepsilon$;
(6) *the function* $\mu \to \int D(q(f))\, d\mu$ *is weak* continuous on space of shift invariant measures* μ *on the Polish space* F.

Note: an element of G *regarded as an element of* $\mathscr{B}$ *acts by left multiplication.*

The complex form of conditions (3) and (4) arise from the following considerations. When an orbit is viewed as a copy of G the base point x sits on the identity element of the group. When acted on by some rearrangement the identity moves, i.e. the point x now is based at a different point in G. Hence it is necessary to view both q^{-1} and q as based at this new origin when they act. Writing it out explicitly on an orbit we have the identities

$$q_x^{\alpha,\psi_2\circ\psi_1} = q_x^{\alpha,\psi_1}(\mathrm{id})q_x^{\alpha,\psi_2}(q_x^{\alpha,\psi_1}(\mathrm{id}))^{-1}q_x^{\alpha,\psi_1}$$

and

$$q_x^{\alpha\psi,\psi^{-1}} = q_x^{\alpha,\psi}(\mathrm{id})(q_x^{\alpha,\psi})^{-1}q_x^{\alpha,\psi}(\mathrm{id})$$

and now conditions (3) and (4) become

$$D(q_x^{\alpha,\psi_2\circ\psi_1}) \le D(q_x^{\alpha,\psi_1}) + D(q_x^{\alpha\psi_1,\psi_1^{-1}\psi_2})$$

and

$$D(q_x^{\alpha,\psi}) = D(q_x^{\alpha\psi,\psi^{-1}}).$$

For size kernels D, defined solely in terms of $H(q)$, these two become even simpler as

$$H(q_1(\mathrm{id})q_2q_1^{-1}(\mathrm{id})q_1) = H(q_2)H(q_1)$$

and

$$H(q(\mathrm{id})^{-1}q^{-1}q(\mathrm{id})) = H(q)^{-1}.$$

For a size kernel D we define

$$m^D(\alpha,\phi) = \int D(q_x^{\alpha,\phi})\,d\mu(x).$$

We call such an m^D an **integral size.**

In condition (6) on D we could have asked for only upper semi-continuity and still obtained that m^D was a size. All examples though are continuous here so we ask for the stronger condition and obtain a stronger conclusion:

Theorem 2.3.2. *For D a size kernel, m^D is a 3^+ size.*

Proof First note that

$$\begin{aligned} q_x^{\alpha\phi_1,\phi_2}(g) &= \alpha(\phi_1(x), \phi_1\phi_2(T_g^{\alpha\phi_1}(x))) \\ &= \alpha(\phi_1(x), \phi_1\phi_2(\phi_1^{-1}T_g^{\alpha}(\phi_1(x)))) \\ &= q_{\phi_1(x)}^{\alpha,\phi_1\phi_2\phi_1^{-1}}(g) \quad \text{and so} \\ m_\alpha^D(\phi_1,\phi_3) &= \int D(q_x^{\alpha\phi_1,\phi_1^{-1}\phi_3})\,d\mu \\ &= \int D(q_{\phi_1(x)}^{\alpha,\phi_3\phi_1^{-1}})\,d\mu \\ &= \int D(q_{\phi_1(x)}^{\alpha,\phi_3\phi_2^{-1}\phi_2\phi_1^{-1}})\,d\mu \\ &\quad \text{and from the above identity and condition (3),} \\ &\leq \int D(q_{\phi_1(x)}^{\alpha,\phi_3\phi_2^{-1}}) + D(q_{\phi_1(x)}^{\alpha,\phi_2\phi_1^{-1}})\,d\mu \\ &= \int D(q_{\phi_2^{-1}\phi_1(x)}^{\alpha\phi_2,\phi_2^{-1}\phi_3})\,d\mu + \int D(q_x^{\alpha\phi_1,\phi_1^{-1}\phi_2})\,d\mu \\ &= m_\alpha^D(\phi_1,\phi_2) + m_\alpha^D(\phi_2,\phi_3). \end{aligned}$$

Condition (3) gives symmetry as

$$D(q_x^{\alpha\phi_1,\phi_2\phi_1^{-1}}) = D(q_x^{\alpha\phi_2,\phi_1\phi_2^{-1}}) = m_\alpha^D(\phi_2,\phi_1)$$

and m_α^D is a pseudometric on Γ. Axiom 2 of a size follows directly from condition (5). Condition (6) is precisely that Axiom 3 should hold. □

Examples 1 & 2 (Conjugacy and Orbit Equivalence). These first two examples are the extremes of what is possible. For one the equivalence class will simply be the full-group orbit and for the other it will be the entire set of arrangements. Both of the pseudometrics $d(q, \mathrm{id})$ and $d(H(q), \mathrm{id})$ are easily seen to be size kernels and so both

$$\begin{aligned} m^1(\alpha,\phi) &= \|(\alpha,\phi),(\alpha,\mathrm{id})\|_\alpha^s \text{ and} \\ m^0(\alpha,\phi) &= \|(\alpha,\phi),(\alpha,\mathrm{id})\|_\alpha^w \end{aligned}$$

are 3^+ sizes.

As d makes $\mathscr{B}$ complete, relative to m^1 a class of sequences $\langle\phi_i\rangle_\alpha \in \hat{\Gamma}_\alpha$ iff $\phi_i \to \phi$ in probability. Thus $\alpha \overset{m^1}{\sim} \beta$ iff $\beta = \alpha\phi$ i.e. they differ by an element of the full-group and the equivalence class of α is exactly its full-group orbit. Note in particular that T^α and T^β will be conjugate actions. To tie this relation into our work here notice that in Chapter 7, where we define the notion of an m-finitely determined action, this definition reduces to Ornstein's classical characterization of the Bernoulli actions as the finitely determined actions for m^1.

As for m^0, for any α and β one can use the Ornstein–Weiss Rokhlin Lemma (Lemma 2.1.5) to construct a sequence of ϕ_i with $\alpha\phi_i \to \beta$ in L^1 with the sequence ϕ_i an m_α-Cauchy sequence. Thus all arrangements are m^0-equivalent. Dye's Theorem [10] and the Theorem of Connes, Feldman and Weiss [6] now tell us that any two ergodic actions of G are m^0-equivalent.

We tie this example into our work. First a reminder of the distribution topology. For G a countable and amenable group and Σ a finite labeling set, the space of probability measures on Σ^G forms a compact metrizable space. For any measure-preserving action of G and Σ valued partition the map from points to Σ, G-names will project the invariant measure to a measure on Σ^G and will make a pseudometric space of such processes (pairs of actions and partitions). For two such pairs (T,P) and (S,Q) let $\|(T,P),(S,Q)\|_*$ be some metric giving this weak* or distribution pseudotopology. We state a lemma concerning ergodic actions of G.

Lemma 2.3.3. *Let G be a countable and amenable group, T^α a free and ergodic action of G on the standard space $(X,\mathcal{F},\mu)$ and P a finite Σ-valued partition of X. For each $\varepsilon > 0$ there is a $\delta > 0$ so that for any other free and ergodic action S^β of G on $(Y,\mathcal{G},\nu)$ and partition $Q : Y \to \Sigma$ satisfying:*
(1) $\|(T^\alpha,P),(S^\beta,Q)\|_* < \delta$

and for every $\delta_1 < 0$ there is a ϕ in the full-group of β and a partition Q' of Y with

(*a*) $m^0(\beta,\phi) < \varepsilon$,

(*b*) $\nu(Q\triangle Q') < \varepsilon$ *and*

(1′) $\|(T^\alpha,P),(S^{\beta\phi},Q')\|_* < \delta_1$.

We leave the proof as an exercise for the reader. A version of this fact for Kakutani Equivalence is found in lemma 4.3 of [38]. The reader can use this as an outline of to how to proceed. One concludes from this that all ergodic actions of G are weakly m^0-finitely determined (see Definition 7.2.5) and, as m^0 is a 3^+ size, Theorem 7.2.6 now implies all ergodic actions are m^0-f.d. giving an alternate albeit elaborate proof of Dye's theorem using our machinery.

Before we continue to other examples we make a few general observations concerning size kernels. First we can w.l.o.g. assume that all size kernels are bounded by 1, as replacing D by the supremum of D and 1 will maintain the axioms and will not change the associated equivalence relation. Notice next that one evaluates the size m^D of a rearrangement by calculating $\int D(Q(f))\,d\mu(f)$ where μ is some shift invariant measure on F. Suppose $F_0 \subseteq F$ is a shift invariant set with $\mu(F_0) = 1$ for all shift invariant probability measures μ. Assume as well that F_0 contains the identity and if it contains $F(q)$ then it also contains $F(q^{-1})$. Notice that changing D outside of F_0 will have no effect on the evaluation of m^D. In particular if D is initially only defined on F_0, is bounded by 1 there and satisfies the axioms of a size kernel (where applicable), then if we set $D(Q(f)) = 1$ for $f \notin F_0$ we would extend D so as to be a size kernel.

We now give an explicit example of such an F_0. Suppose G is $\mathbb{Z}^n$ and $B_N = [-N, N]^n$ is the standard Følner sequence of boxes. For shift invariant measures on G^G the pointwise ergodic theorem holds along this sequence B_N. For each $f \in F$ set $\Delta_M(f)$ to be the upper density of the set $\{\vec{v} | f(\vec{v}) \notin B_M\}$ calculated along the sequence of sets B_N as $N \nearrow \infty$. Let F_0 consist of those f for which $\lim_{M\to\infty} \Delta_M(f) = 0$. The pointwise ergodic theorem tells us that this set has measure 1 for all shift invariant measures. Hence when working in $\mathbb{Z}^n$ one need only define a size kernel D on such f. Notice that for such f one will have

$$\lim_{N\to\infty} \frac{\#\{\vec{v} \in B_N | Q(f)(v) \notin B_N\}}{\#B_N} = 0.$$

We describe a class of examples that take advantage of these observations and this choice for an F_0.

Example 3 (Kakutani equivalence). The development of Kakutani equivalence in $\mathbb{Z}^n$ can be found in [9] and a complete development of the equivalence theorem for it in [16]. What we present here is an approach that brings this example into our context. For this example let $G = \mathbb{Z}^n$ and $B_N = [-N, N]^n$ be the standard Følner sequence of boxes centered at $\vec{0}$. We begin with a metric on $\mathbb{Z}^n$ given by

$$\tau(\vec{u}, \vec{v}) = \min\big(\|(\vec{u}/\|\vec{u}\|) - (\vec{v}/\|\vec{v}\|)\| + |\ln(\|\vec{v}\|) - \ln(\|\vec{u}\|)|, 1\big)$$

(assuming $\vec{0}/\|\vec{0}\| = \vec{0}$). What is important about τ are the following two properties:

(1) τ is a metric on $\mathbb{Z}^n$ bounded by 1; and

(2) $\vec{u}$ and $\vec{v}$ are τ close iff the norm of their difference is small in proportion to both of their norms.

For $h \in \mathcal{G}$ set $B_N(h) = \{\vec{v} \in B_N | h(\vec{v}) \in B_N\}$ (those elements of B_N mapped into B_N by h). Now set

$$k(h) = \sup_N \left(\frac{1}{\# B_N} \left(\sum_{\vec{v} \in B_N(h)} \tau(\vec{v}, h(\vec{v})) + \#\{\vec{v} \in B_N | h(\vec{v}) \notin B_N\} \right) \right)$$

and $K(q) = k(H(q))$.

Lemma 2.3.4. *The function K is a size kernel.*

Proof That $k(h) = k(h^{-1})$ is a calculation as in fact this equality holds already for each N. That $k(h_2 \circ h_1) \leq k(h_2) + k(h_1)$ is also true as it is true for each N before taking the $\sup_N$. That K satisfies the first five conditions of a size kernel is now direct. We get (3) and (4) from the observation that $K(q)$ only depends on $H(q)$. As τ is a metric, for $K(q)$ to be small $H(q)$ must fix a large (finite of course) number of vectors $\vec{v}$. To obtain (6) suppose μ is some shift invariant measure on G^G, hence supported on F_0. For $q \in F_0$ as $N \to \infty$ we have $|B_N(H(q))|/|B_N| \to 1$. Moreover for each N this value is continuous and so its expected value relative to μ is weak* continuous. For any fixed $\vec{v}$, $\{q | H(q)(\text{id}) = \vec{v}\}$ is a clopen set and hence its measure is weak* continuous in μ. As $\vec{v}$ varies these sets form a countable partition of G^G and so for any $\varepsilon > 0$ there is an N_0 and a neighborhood U of μ so that for $\nu \in U$ also invariant and $N \geq N_0$, letting $h = h(q(f))$,

$$\int \frac{1}{\# B_N} \left(\sum_{\vec{v} \in B_N(h)} \tau(\vec{v}, h(\vec{v})) + \#\{\vec{v} \in B_N | h(\vec{v}) \notin B_N\} \right) d\mu(f) < \varepsilon.$$

For each $N < N_0$ the calculation

$$\frac{1}{\# B_N} \left(\sum_{\vec{v} \in B_N(h)} \tau(\vec{v}, h(\vec{v})) + \#\{\vec{v} \in B_N | h(\vec{v}) \notin B_N\} \right)$$

is continuous and hence the supremum of these values for $N < N_0$ is continuous and so its integral is weak* continuous in μ. It follows that in some sub-neighborhood $U' \subseteq U$ we will have a variation of at most ε in the value $\int K(q(f))\, d\nu$. □

The use of τ here is not the usual calculation taken to construct Kakutani equivalence, but noting that for τ to be small simply means the distance between two vectors is small relative to their lengths makes it clear that it is equivalent to earlier presentations. One finds without much effort that $\alpha \overset{m^K}{\sim} \beta$ iff for a.e. x

$$\limsup_{N\to\infty} \frac{1}{\#B_N} \sum_{\vec{v}\in B_N} \tau(h_x^{\alpha,\beta}(\vec{v}),\vec{v}) = 0.$$

Although the vocabulary of [9] is somewhat different it is shown there that this is equivalent to saying:

Proposition 2.3.5. *For every $\varepsilon > 0$ there is a $\phi \in \Gamma$ and a subset A with $\mu(A) > 1-\varepsilon$ so that for all $x, y \in A$,*

$$\tau(\alpha\phi(x,y), \alpha(x,y)) < \varepsilon.$$

For $G = \mathbb{Z}$ this implies that $T^{\alpha\phi}$ and T^{β} induce the same map on A and hence T^{α} and T^{β} are evenly Kakutani equivalent in the classical sense. In [9] the converse of this is proven, i.e. this is precisely even Kakutani equivalence in $\mathbb{Z}$ and a broad exploration of this equivalence relation in $\mathbb{Z}^d$ is made connecting it to Katok cross-sections of $\mathbb{R}^d$ actions.

As m^K is entropy-preserving we know the Bernoulli actions are m^K-finitely determined and hence there exist m^K-finitely determined actions. By Theorem 7.2.6 they are characterized by the condition of being weakly m^K-finitely determined. Notice that for actions of $\mathbb{Z}$ this precise fact is proven in lemma 4.3 of [38].

Examples 4 & 5 (α equivalences). Once more take $G = \mathbb{Z}^n$ and choose a vector $\vec{\alpha} = \{\alpha_1, \alpha_2, \ldots, \alpha_n\}$ of nonzero real numbers. Set $A : \mathbb{R}^n \to \mathbb{T}^n$ to be

$$A(v_1, \ldots, v_n) = (v_1/\alpha_1, v_2/\alpha_2, \ldots, v_n/\alpha_n) \mod \mathbb{Z}^n$$

and on $\mathbb{T}^n$ use the natural metric

$$\rho(\vec{v}_1, \vec{v}_2) = \|e^{2\pi i \vec{v}_1 \cdot \vec{1}} - e^{2\pi i \vec{v}_2 \cdot \vec{1}}\|.$$

Notice that for $\rho \circ A$ to be small means the two vectors differ approximately by a vector $\{n_1\alpha_1, \ldots, n_i\alpha_i\}$ where the n_i are integers.

For $q \in \mathscr{B}$ set $\mathbb{A}(q) = \rho(A(q(\vec{0})), A(\vec{0}))$. $\mathbb{A}$ is not a size kernel as it fails to satisfy (1) although it does satisfy both (2) and (3). To obtain a size kernel all we need do is add to $\mathbb{A}$ some other size kernel. We currently have two choices giving the two size kernels

$$\begin{aligned} D_{\vec{\alpha}}(q) &= d(H(q), \mathrm{id}) + \mathbb{A}(q) \quad \text{and} \\ K_{\vec{\alpha}}(q) &= K(q) + \mathbb{A}(q). \end{aligned}$$

(Notice it makes no sense to use $d(q, \mathrm{id})$ as adding on $\mathbb{A}(q)$ would add no further restriction to the already minimal equivalence class.)

Both of these examples give interesting equivalence relations. The second has been well studied under the name of α-equivalence (see [8] and [48]). Because of the standard use of $\vec{\alpha}$ to represent the parameter of this relation we will use β to represent an arrangement throughout the discussion of these two examples.

The first example, $D_{\vec{\alpha}}$, has not been discussed in the literature so we present a brief discussion here. What we obtain is a refinement of simple orbit equivalence that splits the ergodic actions into a countable list of equivalence classes characterized spectrally.

Remember a function $f : X \to \mathbb{C}$ is an eigenfunction of the ergodic action T with eigenvalue $\vec{\lambda}$ if f is of norm one and

$$f \circ T_{\vec{v}} = e^{2\pi i \vec{v}\cdot\vec{\lambda}} f.$$

Fixing $\vec{\lambda}_0 = (1/\alpha_1, 1/\alpha_2, \ldots, 1/\alpha_n)$, those values $(k_1, k_2, \ldots, k_n) \in \mathbb{Z}^n$ for which $(k_1/\alpha_1, k_2/\alpha_2, \ldots, k_n/\alpha_n)$ is an eigenvalue for T form an additive subgroup we will call $\Lambda_{\vec{\alpha}}(T)$. Two ergodic actions U and V of $\mathbb{Z}^n$ are $m^{D_{\vec{\alpha}}}$ equivalent iff $\Lambda_{\vec{\alpha}}(U) = \Lambda_{\vec{\alpha}}(V)$. In particular we see that there are at most countably many $m^{D_{\vec{\alpha}}}$-equivalence classes. We will indicate the proof of parts of this characterization, leaving much to the reader.

Proposition 2.3.6. *If* $\beta_1 \overset{m^{D_{\vec{\alpha}}}}{\sim} \beta_2$ *then* $\Lambda_{\vec{\alpha}}(T^{\beta_1}) = \Lambda_{\vec{\alpha}}(T^{\beta_2})$.

Proof Suppose $\vec{\lambda} = (k_1/\alpha_1, k_2/\alpha_2, \ldots, k_n/\alpha_n)$ is an eigenvalue for the eigenfunction f of T^{β_1}. We now compute that for all $\vec{v}$

$$\begin{aligned}
\|f \circ T_{\vec{v}}^{\beta_1\phi} - e^{2\pi i\vec{v}\cdot\vec{\lambda}} f\|_1 &= \|f \circ T_{\vec{v}}^{\beta_1\phi} - f \circ T_{\vec{v}}^{\beta_1}\|_1 \\
&= \|f \circ T^{\beta_1}_{\vec{v}+\beta_1(x,\phi(x))-\beta_1(T_{\vec{v}}^{\beta_1}(x),\phi(T_{\vec{v}}^{\beta_1}(x)))} - f \circ T_{\vec{v}}^{\beta_1}\|_1 \\
&= \|f \circ T^{\beta_1}_{\beta_1(x,\phi(x))-\beta_1(T_{\vec{v}}^{\beta_1}(x),\phi(T_{\vec{v}}^{\beta_1}(x)))} - f\|_1 \\
&= \|e^{2\pi i(\beta_1(x,\phi(x))-\beta_1(T_{\vec{v}}^{\beta_1}(x),\phi(T_{\vec{v}}^{\beta_1}(x))))} f - f\|_1 \\
&\le \|\rho(A(q_x^{\beta_1,\phi}(\vec{0})), A(\vec{0}))\|_1 + \|\rho(A(q^{\beta_1,\phi}_{T_{\vec{v}}^{\beta_1}(x)}(\vec{0})), A(\vec{0}))\|_1 \\
&\le 2m^{D_{\vec{\alpha}}}(\beta_1, \phi).
\end{aligned}$$

Thus if $\beta_1 \overset{m^{D_{\vec{\alpha}}}}{\sim} \beta_2$, for all $\vec{v}$ we will have

$$\|f \circ T_{\vec{v}}^{\beta_2} e^{-2\pi i\vec{v}\cdot\vec{\lambda}} - f\|_1 \le 2m^{D_{\vec{\alpha}}}(\beta_1, \beta_2).$$

By the mean ergodic theorem there must be an $\hat{f}$ with

$$\frac{1}{\#B_N}\sum_{\vec{v}\in B_N} f\circ T_{\vec{v}}^{\beta_2}e^{-2\pi i\vec{v}\cdot\vec{\lambda}} \longrightarrow \hat{f}$$

and by the above,

$$\|f-\hat{f}\|_1 \leq 2m^{D_{\vec{\alpha}}}(\beta_1,\beta_2).$$

We know $\hat{f}$ must have constant norm and as long as it is not identically 0, $\hat{f}/|\hat{f}|$ will be an eigenfunction for T^{β_1} with eigenvalue $\vec{\lambda}$. As all $T^{\beta_1\phi}$ are conjugate to T^{β_1} we can assume $m^{D_{\vec{\alpha}}}(\beta_1,\beta_2) < 1/2$, forcing $\hat{f}\neq 0$. □

Lemma 2.3.7. *Suppose T^β is a free and ergodic action of $\mathbb{Z}^n$ on $(X,\mathscr{F},\mu)$ with $\Lambda_{\vec{\alpha}}(T^\beta)=\{\vec{0}\}$, and $P : X\to\Sigma$ is a finite partition. For each $\varepsilon>0$ there is a $\delta>0$ so that for any other free and ergodic action S^γ on $(Y,\mathscr{G},\nu)$ and partition $Q : Y\to\Sigma$ satisfying:*

(1) $\|(T^\beta,P),(S^\gamma,Q)\|_* < \delta$

and any value $\delta_1<0$ there is a ϕ in the full-group of S^γ and a partition $Q' : Y\to\Sigma$ with

(*a*) $m^{D_{\vec{\alpha}}}(\gamma,\phi)<\varepsilon$

(*b*) $\nu(Q\triangle Q')<\varepsilon$ *and*

(1′) $\|(T^\beta,P),(S^{\gamma\phi},Q')\|_* < \delta_1$.

We once more leave a complete proof of this copying lemma to the reader. We do point out the ingredient used to obtain (a) beyond the construction of Lemma 2.2.3. For $\vec{\alpha}$ fixed consider the group rotation on $\mathbb{T}^n$ given by $(x_1,\ldots,x_n)\to(x_1+\alpha_1,\ldots,x_n+\alpha_n)$ mod 1. This is not necessarily ergodic but all its ergodic components are conjugate to some group rotation we call $(R_{\vec{\alpha}},Z)$ where Z is a compact subgroup of $\mathbb{T}^n$. To say $\Lambda_{\vec{\alpha}}(T^\beta)=\{\vec{0}\}$ is equivalent to saying $R_{\vec{\alpha}}\times T^\beta$ is ergodic. Partition Z into sets of diameter less than $\varepsilon/2$ by a partition H.

Consider now $H\vee P, B_N$-names arising from the action of $R_{\vec{\alpha}}\times T^\beta$. The pointwise ergodic theorem tells us that if we fix a choice of $h\in H$ and cylinder set C in the process (T^β,P) then for N large the relative density of C just at indices of an $H\vee P, B_N$-name whose H term is h will be very close to $\mu(C)$.

Fixing the value N, if (S^γ,Q) satisfies (1) for a small enough δ then this same fact will be true for $H\vee Q$-names relative to ν (even if $R_{\vec{\alpha}}\times S^\gamma$ is not ergodic). The full-group element ϕ will now be constructed on some Rokhlin tower of size B_M, $(M >> N)$ in the action S^γ by overlaying the S^γ, Q names with a template $R_{\vec{\alpha}}, H$-name and constructing ϕ to move this name close to some T^β, P-name in $\bar{d}$ while simultaneously preserving the template name. The remark in the previous paragraph guarantees that this can be done. This preservation of the template name is the new ingredient needed to ensure $m^{D_{\vec{\alpha}}}(\gamma, \phi) < \varepsilon$. One now changes Q slightly according to the necessary $\bar{d}$ error to be an exact copy of the target T^β, P-name.

This Lemma and the previous Proposition now imply that the $m^{D_{\vec{\alpha}}}$-finitely determined actions are those with $\Lambda_{\vec{\alpha}}(T^\beta) = \{\vec{0}\}$ (see Definition 7.2.5 and Theorem 7.2.6) and any two such actions will be $m^{D_{\vec{\alpha}}}$-equivalent.

One approach to showing that the group $\Lambda_{\vec{\alpha}}(T^\beta)$ is a complete invariant even outside the finitely determined class is to fix a choice Λ for $\Lambda_{\vec{\alpha}}$ and generalize Lemma 2.3.7 and the theory of $m^{D_{\vec{\alpha}}}$-joinings of Chapter 6 to a relativized version relative to the non-trivial isometric factor algebra generated by the eigenfunctions whose eigenvalues lie in Λ.

As indicated earlier the size kernel $K_{\vec{\alpha}}$ leads to a rather well studied area. Certainly $m^{K_{\vec{\alpha}}}$-equivalence refines even Kakutani equivalence. The argument in Proposition 2.3.6 applies here as well to say the group $\Lambda_{\vec{\alpha}}$ is an $m^{K_{\vec{\alpha}}}$ invariant. Hence each even Kakutani equivalence class is cut into at least countably many $m^{K_{\vec{\alpha}}}$ classes. In [8] it is shown that in one dimension, and for the "loosely Bernoulli" class (the m^K-finitely determined class), this is the full refinement. That argument will push through to all dimensions. As a consequence of the construction in [15] there are other m^K classes which contain uncountably infinitely many $m^{K_{\vec{\alpha}}}$ classes.

There exists a connection to sections for actions of $\mathbb{R}^n$ as well, although it is understood only in 1 and 2 dimensions. The following is what is known in one dimension. For α irrational and > 0 any measure-preserving flow can be represented as a flow under a function taking on only the two values 1 and $1 + \alpha$ [45]. Just as for Kakutani equivalence itself one can define here a relationship between $\mathbb{Z}$ actions by saying they are α-related if they arise as the return maps from a common flow where the return times take on only these two values 1 and $1 + \alpha$. We say they are evenly α-related iff the integrals of these return time functions agree. In [8] it is shown that two $\mathbb{Z}$ actions are α-related iff they are $m^{K_{\vec{\alpha}}}$-equivalent. Moreover it is shown that if U and V are $m^{K_{\vec{\alpha}}}$-equivalent, then any flow for which U arises as such a section, V does as well.

Here is what is known in two and higher dimensions. In [47] it is shown that any $\mathbb{R}^n$ action can be represented as a special sort of "Markov" tiling suspension of an action of $\mathbb{Z}^n$ where the tiles are rectangles whose length in dimension k is either only 1 or $1+\alpha_k$ (we assume all α_k are irrational and > 0). We say two actions U and V of $\mathbb{Z}^n$ are $\vec{\alpha}$ related if they arise as such sections of a common $\mathbb{R}^n$ action. We say they are evenly $\vec{\alpha}$-equivalent if the proportion of the space occupied by each of the tile shapes is the same for both representations. The argument in [8] extends to higher dimensions to show that if U and V are evenly $\vec{\alpha}$-related then they are $m^{K_{\vec{\alpha}}}$-equivalent. In two dimensions Sahin [48] shows that the converse is true, i.e. if two actions U and V are $m^{K_{\vec{\alpha}}}$-equivalent then they arise as Markov tiling sections of a common $\mathbb{R}^n$ action. It remains open however whether any $\mathbb{R}^n$ action for which U is such a section must also have V as such a section.

Our last two examples exhibit another general context in which a restricted orbit equivalence relation can arise. Suppose that to each arrangment α we can assign a subgroup Γ_0^α of the full-group Γ with the equivariance property that $\Gamma_0^{\alpha\phi} = \phi^{-1}\Gamma_0^\alpha\phi$. In particular the choice of subgroup does not change when we perturb α by an element of its subgroup. What interests us are those β reachable as limits of sequences of rearrangements $\alpha\phi_i$ where $\phi_i \in \Gamma_0^\alpha$. We write a size for such a relation as a sum of two pieces, one measuring the m_α^1 distance from ϕ to Γ_0^α and the other measuring some chosen size of (α, ϕ). As a simple example of this consider α-equivalence in $\mathbb{Z}$ for $\alpha = 2$. Here the groups $\Gamma_0^\alpha = \{\phi | \alpha(x, \phi(x)) = 0 \mod 2\}$. Using the two sizes m^0 and m^K within the subgroups will yield the two examples described of $\vec{\alpha}$-equivalence. This idea becomes cumbersome for α_i irrational. We do not attempt to axiomatize precisely what is needed of the family of subgroups, leaving this to a more general study of sizes. Our final two examples will offer an indication of the range this idea covers.

Example 6 (Vershik's $\vec{r}$ Equivalence). The work described here can be found in [18] and [19] of Heicklen. Suppose $(X, \mathscr{F}, \mu)$ is a standard non-atomic probability space and $\mathscr{F}_i$ is a sequence of sub σ-algebras with $\mathscr{F} = \mathscr{F}_0$ and $\mathscr{F}_{i+1} \subseteq \mathscr{F}_i$. We refer to such a sequence as a **reverse filtration**. Two such sequences are **conjugate** if there is a measure-preserving bijection between the measure spaces carrying one reverse filtration, term by term, to the other. To remove some trivial issues and make this subject addressable by our methods we make two assumptions. First, we assume that for each i the conditional fiber measures of $\mathscr{F}_i$ over

$\mathscr{F}_{i-1}$ are atomic with a fixed number of atoms k_i and that each atom has a constant mass $1/k_i$. We call such a filtration **uniform**. Next we assume that the $\mathscr{F}_i$ decrease to the trivial algebra. A filtration with this property is called **exact**. One natural way for such a filtration to arise is from an action of a group of the form $G = \sum_{n=1}^{\infty} \mathbb{Z}/r_n\mathbb{Z}$. What matters here is that G is the increasing union of the finite groups $H_i = \sum_{n=1}^{i} \mathbb{Z}/r_n\mathbb{Z}$. If we have a measure-preserving and free action of this group and we set $\mathscr{F}_i$ to be the algebra of H_i invariant sets then we obtain a uniform reverse filtration. It is exact iff the action is ergodic. Conversely, given any uniform and exact reverse filtration, using the Rokhlin decomposition of each successive $\mathscr{F}_i$ over $\mathscr{F}_{i-1}$, we can place on the space an action of G for which the filtration is obtained as this list of invariant sub-algebras. The action of G here is not unique and this leads to a natural relation: we say two actions of G are Vershik related if conjugate versions of both of them can be placed on the same space, giving rise to the same reverse filtration. Notice in particular that the two actions will be orbit-equivalent and what characterizes the particular orbit equivalence is that it preserves the orbits of all the subgroups H_i.

For the purpose of our discussion it will be useful to assume only that G is the increasing union of finite abelian groups H_i without assuming that each is cyclic over its predecessor. Such a G is countable and amenable. Notice that for any increasing subsequence $\{j_i\}$ we could define $\hat{H}_i = H_{j_i}$ and get another representation of G as an increasing union of finite subgroups. These distinct representations will give distinct values for the vector $\vec{r} = \{|\hat{H}_1|, |\hat{H}_2/\hat{H}_1|, \dots\}$ and so we can represent a choice for such an increasing subsequence of subgroups by its vector $\vec{r}$. This is consistent with the usage when H_i/H_{i-1} is cyclic of order r_i. We say two actions of G are Vershik $\vec{r}$-related if they are Vershik related for the choice of subgroups $\hat{H}_i$ determined by the values of $\vec{r}$. We describe Vershik relatedness indexed by the choice of subsequence $\vec{r}$ as a family of restricted orbit equivalences on G.

Although one can use a size kernel here we follow [19] and give the size directly. Notice that for a fixed arrangement α and choice for $\vec{r}$ the full-group Γ contains closed subgroups $\Gamma_0^{\vec{r},\alpha}$ consisting of those ϕ which preserve the T^α orbits of all $\hat{H}_i$. (This is equivalent to saying that either $q_x^{\alpha,\phi}$ or equivalently $h_x^{\alpha,\phi}$ permutes cosets of $\hat{H}_i$ for all i and a.e. x.) If we have a sequence of rearrangements $\alpha\phi_i$ converging to some β where all the $\phi_i \in \Gamma_0^{\vec{r},\alpha}$ then T^α and T^β will have identical $\hat{H}_i$ orbits for all i and hence be Vershik $\vec{r}$ related in this very strong sense. To define a size giving this relation, for α and $\vec{r}$ fixed, we first calculate the distance some

ϕ is from the subgroup $\Gamma^{\vec{r},\alpha}$ as its m^1_α distance:

$$c_{\vec{r}}(\alpha,\phi) = \inf_{\phi' \in \Gamma_0^{\vec{r},\alpha}} \mu\{x|\phi(x) \neq \phi'(x)\}.$$

(This is not the definition of c_r given in [19] but is equivalent by the **Flattening Lemma** proven there.) One can now define the family of sizes

$$m^{\vec{r}}(\alpha,\phi) = c_{\vec{r}}(\alpha,\phi) + m^0(\alpha,\phi).$$

Heicklen proves this to be a size but does not show it to be 3^+. We will not present the details showing that it is in fact 3^+. This can be done either by suitably expanding Heicklen's argument or by showing that the sizes $m^{\vec{r}}$ arise from size kernels.

Heicklen's conclusion is that two actions $\alpha \overset{m^{\vec{r}}}{\sim} \beta$ iff there is a $\psi \in \Gamma$ so that $T^{\alpha\psi}$ and T^β have identical $\hat{H}_i$ orbits for all i. As T^α and $T^{\alpha\psi}$ are conjugate (by ψ of course) T^α and T^β are $\vec{r}$ related and if two actions are $\vec{r}$ related they can be realized as two such actions.

The family of sizes exhibits two very interesting properties. The first is due to Vershik who proved a **lacunary isomorphism theorem** for such groups [58]: for any two actions U and V, if the r_i are chosen to grow rapidly enough, then the two actions are $\vec{r}$ related. Notice that the allowed choices for the sequence $\hat{H}_i$ are partially ordered under containment. As one goes *further out* in this partial order more and more actions become equivalent and Vershik's result says any two will become equivalent once one is far enough out in this net.

The second very interesting property follows from this. Vershik has shown that for very slowly growing sequences, like $r_i = 2$ for all i, the entropy of a G-action is an invariant of $\vec{r}$-equivalence. On the other hand, Vershik's lacunary isomorphism theorem tells us that for some choices of $\vec{r}$, entropy definitely is not an invariant. As we learn in Chapter 5 a restricted orbit equivalence either preserves entropy or generically in the m_α topology an action has zero entropy. Vershik [59] conjectured and proved the sufficiency and Heicklen [19] proved the necessity of the following characterization of the boundary between these two regimes: The size $m^{\vec{r}}$ is entropy-preserving iff

$$\sum_{i=1}^{\infty} \frac{\log r_{k+1}}{\#H_k} < \infty.$$

Example 7 (Entropy as a Size). We discuss this example only for actions of $\mathbb{Z}$ although the ideas extend to general countable amenable groups. The results described here are found in [46]. Before examining this example in detail consider the following observations. Two major goals of this current work are to demonstrate:

(1) A size m is either entropy-preserving in that two equivalent actions have the same entropy, or entropy-free in that residually in each class actions have zero entropy. In the first case we say an action's m-entropy is its entropy and in the latter that its m-entropy is always zero.
(2) Each size possesses a family of distinguished classes, characterized by their m-entropy, called the m-finitely determined classes. Any two m-finitely determined actions of the same m-entropy are m-equivalent.

Notice that this implies the possibility of two sizes m for which all actions are finitely determined, one that is entropy-free and one that is entropy-preserving. Dye's theorem, here done via the size m^0, shows that there is an entropy-free size for which all actions are m-finitely determined. What the example we now discuss shows is that the other size also exists, relative to which two actions are equivalent iff they have the same entropy.

The size at its base will simply be the entropy of the rearrangment itself. We make this precise as follows. Note $g_{(\alpha,\phi)}(x) = \alpha(x, \phi(x))$ takes on countably many values and hence can be regarded as a countable partition $g_{(\alpha,\phi)}$ of X. Set Γ_0^α to be those ϕ for which $g_{(\alpha,\phi)}$ is finite. It is not difficult to see that Γ_0^α is a subgroup and moreover $\Gamma_0^{\alpha\psi} = \psi^{-1}\Gamma_0^\alpha\psi$ as $g_{(\alpha\psi,\psi^{-1}\phi\psi)}(\psi^{-1}(x)) = g_{(\alpha,\phi)}(x)$. It is shown in Theorem 4.0.2 that the Γ_0^α are all m_α^1 dense in Γ. For $\phi \in \Gamma_0^\alpha$ one can use the entropy of the process $h(T^\alpha, g_{(\alpha,\phi)})$ to start the definition of a size defining

$$e(\alpha, \phi) = \inf_{\phi' \in \Gamma_0^\alpha} h(T^\alpha, g_{(\alpha,\phi')}) + \mu\{x | \phi(x) \neq \phi'(x)\}.$$

Now set the size to be

$$m^e(\alpha, \phi) = e(\alpha, \phi) + m^0(\alpha, \phi).$$

To see that this is a size, Axiom 1 follows from basic conditional entropy considerations and Axiom 2 is directly due to the second term. Axiom 3 here follows from upper semi-continuity of entropy and for this reason this is not on the face of it a 3^+ size. To see that in fact it is not, note that those α for which T^α is zero entropy are m^0 residual. Hence a rearrangment (α, ϕ) can be perturbed by as little as we like in distribution to an (α', ϕ') with $m^e(\alpha', \phi') = 0$. It is this example which motivated the weakening of Axiom 3^+ to Axiom 3.

The form of the size makes it reasonable to believe and easy to prove that m^e-equivalence will be entropy-preserving. A more subtle combinatorial argument leads to the reverse conclusion as well, that any two ergodic actions of equal entropy are in fact m^e-equivalent.

A Non-Example. Hoffman and Rudolph [20] have presented an isomorphism theory for measure-preserving endomorphisms that can be viewed as an extension of the methods here but cannot be viewed as an application of restricted orbit equivalence as we develop it. This work grows naturally from the Vershik equivalence theory described in Example 6. Consider the standard example discussed there, $\{0, 1, \ldots, p-1\}^{\mathbb{N}}$ with uniform Bernoulli product measure $(1/p, \ldots, 1/p)$. In Example 6 this gives the standard filtration where all $k_i = p$. Here we consider in addition the shift map, giving a $p-1$ Bernoulli endomorphism. We call it "uniform" as all p inverse images of a point are equally likely.

An endomorphism conjugate to this standard one would also have to be uniformly $p-1$ in that almost every point must have p inverse images and all must be equally likely. This standard example has entropy $\log p$ so this also would be true of any endomorphism conjugage to it. What is shown in [20] is that for the class of uniformly $p-1$ endomorphisms of entropy $\log p$ there is an isomorphism theory completely analogous to that of Ornstein for Bernoulli automorphisms and following the same outline as our work here. We will describe enough of [20] to indicate why the theory is parallel and why the results here simply do not apply.

Let T acting on $(X, \mathscr{F}, \mu)$ be an ergodic and uniformly $p-1$ endomorphism of entropy $\log p$. For a.e. $x \in X$ we can consider the set of all inverse images $T^{-j}(x)$, $x > 0$ organized as a p-ary tree rooted at x. The points in $T^{-j}(x)$ are at "level j" of the tree and each $x_1 \in T^{-j}(x)$ is connected by one edge to $T(x_1) \in T^{-j+1}(x)$. We refer to a map from the nodes of such a tree to itself that preserves the edges as a "tree automorphism" and between two such trees as a "tree isomorphism". A conjugacy between two uniformly $p-1$ endomorphisms will give tree isomorphisms between the trees of inverse images attached to matched points.

For P a finite partition of X label each node of a tree of inverse images by the set in P containing it. Given two such labeled trees of inverse images one can ask how closely they can be matched by a tree isomorphism. To be more precise, each node at level j is given a mass of 2^{-j} so that the set of nodes at each level has total mass 1. Relative to this weighting one seeks to match the nodes of two labeled trees by a tree isomorphism minimizing the proportion of nodes in the trees with mismatched labels. This minimum is called the $\bar{t}$ distance between the two labeled trees. This extends in a standard fashion to provide a $\bar{t}$ distance between measures on labeled trees and hence between uniformly $p-1$ P-valued stationary stochastic processes. Notions of $\bar{t}$ finitely-determined and $\bar{t}$ very weakly Bernoulli follow directly and can be shown to be equivalent. Both these are true of the standard example and are conjugacy invariants. Finally, one can give a natural weak* Polish space of joining measures for this theory, called the "one-sided joinings", and show that for the $\bar{t}$ finitely determined endomorphisms those one-sided joinings which arise from conjugacies are a residual subset. This development follows the outline of our work here and is in fact much simpler both because it is just a single example of an equivalence relation and because it is a "zero entropy" theory.

Example 6 of Vershik equivalence was brought under the restricted orbit equivalence umbrella by the choice of an action whose orbit structure mirrored the fibers of the filtration. Here one can also find such a group action. Construct the natural orbit relation setting $x_1 \sim x_2$ if there are $j_1 \geq 0$ and $j_2 \geq 0$ with $T^{j_1}(x_1) = T^{j_2}(x_2)$. The equivalence classes are organized naturally as a complete p-ary tree (each node has $p+1$ edges attached to it, p going back in time and 1 going forward). These naturally organize as the orbits of a free group on p involutions where two points are connected by an edge in the tree iff they are interchanged by one of these p involutions. This is not an amenable group so our work here does not apply. This action also is not measure-preserving. It is in fact an amenable action of type $\mathrm{III}_{1/p}$. As such it could be given as an orbit of a $\mathbb{Z}$ action but this would lose the tree structure essential to the conjugacy theory of the original endomorphisms. What is needed to make this non-example an example is to lift the work here to non-singular amenable actions of groups that are not necessarily amenable. This non-example gives evidence for, and an approach to, such a generalization.

3
The Ornstein–Weiss Machinery

In this section, we describe the constructive tools we will need in order to continue with our work. From the beginnings of the Ornstein approach to constructive ergodic theory and in particular the Isomorphism Theorem it has been understood that there are three basic tools necessary to work constructively with dynamical systems: a version of the Rokhlin lemma; a version of the Ergodic theorem; and a version of the Shannon–McMillan theorem. It has also been understood for some time that a natural context in which all these results hold is that of locally compact and amenable groups. The results described here are lifted almost verbatim from the seminal work of Ornstein and Weiss on this subject [37]. We include them here to provide the reader with ready access to them and as we vary their statements slightly in places. Since we will consider only countable discrete amenable group actions, we do not need the most general form of their results. Thus, for clarity, we have stated these results in the context of discrete group actions. Furthermore, we have opted for a classical description of entropy, using finite partitions and name-counting techniques.

Two notions of essential invariance of finite subsets $F \subseteq G$ are central to [37].

Definition 3.0.1.

Let $\delta > 0$. Let $K \subseteq G$ be a finite set. A subset $F \subseteq G$ is called $(\boldsymbol{\delta}, \mathbf{K})$-**invariant** *if*

$$\frac{\#(KK^{-1}F \triangle F)}{\#F} < \delta.$$

To say that a set F is **sufficiently invariant** *means that there exists a $\delta > 0$ and a finite set $K \subseteq G$ such that F is (δ, K)-invariant.*

To say that a list of sets $F_1, F_2, \ldots, F_k$ is **sufficiently invariant** *is to say that there exists a $\delta > 0$ and a finite set $K \subseteq G$, such that, setting $F_0 = K$, for each $j \in \{1, 2, \ldots, k\}$, the set F_j is (δ, F_{j-1})-invariant.*

We now describe our version of the Ornstein–Weiss quasi-tiling theorem.

Definition 3.0.2.

A finite list of sets $H_1, H_2, \ldots, H_k \subseteq G$, with $id \in H_i$, for all i, is said to **ε-quasi-tile** *a finite set $F \subseteq G$ if there exist "centers" $c_{i,j}$, $i = 1, 2, \ldots, k$, $j = 1, 2, \ldots, l(i)$, and subsets $H_{i,j} \subseteq H_i$ such that:*

(1) $\#H_{i,j} \geq (1 - \varepsilon)\#H_i$, *for* $j = 1, \ldots, l(i)$;

(2) *the* $H_{i,j}c_{i,j} \subseteq F$ *are disjoint; and*

(3) $\#(\bigcup_{i,j} H_{i,j}c_{i,j}) \geq (1 - \varepsilon)\#F$.

Theorem 3.0.3 ([37]). *Given $\varepsilon > 0$, there exists $N = N(\varepsilon)$ such that in any countable discrete amenable group G, if $H_1, \ldots, H_N$ is any sufficiently invariant list of sets, then for any $D \subseteq G$ that is sufficiently invariant (depending on the choice of $H_1, \ldots, H_N$), D can be ε-quasi-tiled by $H_1, \ldots, H_N$.*

This theorem is the essential content of Theorem 6, I.2 [37]. Our definition of ε-quasi-tiling is slightly different; weaker in that we do not ask that $H_i c_{i,j} \cap H_k c_{k,l} = \emptyset, i \neq k$, and stronger in that we require $H_{i,j}c_{i,j} \subseteq F$. Obtaining the latter from Theorem 6, I.2 [37] is easy if D is sufficiently invariant and N is fixed.

We have described this result, as the picture it gives makes much of [37] more accessible. We will not go further into the development of families of tilings, which are the essential tools of their proofs. Rather, we will move on to state their principle results. In particular, we will discuss their version of the Rokhlin and strong Rokhlin lemmas, which can be regarded as dynamical versions of the tiling theorem.

Suppose $(X, \mathscr{B}, \mu)$ is a standard probability space. Suppose T is a measure-preserving free action of G on X. For a finite set $F \subseteq G$ and measurable subset $A \in \mathscr{B}$ with $\mu(A) > 0$, consider $F \times A \subseteq G \times X$. As a measure on $F \times A$, put the direct product $c \times \mu$ of counting measure c and μ. Consider the map $T : F \times A \to X$ given by $T(g, x) = T_g(x)$. On each level set $g \times A$, T is 1-1 and measure-preserving. On any fiber set $F \times x$, T is again 1-1. We definitely do not expect T to be 1-1 on $F \times A$. It is clear, though, that T is non-singular and, at most, $\#F$ to 1.

In particular, if T is j-to-1 at $x \in X$, then

$$\frac{dT^*(c \times \mu)}{d\mu}(x) = j.$$

Rokhlin lemmas concern the degree to which maps T, as above, can be made 1-1. In particular, within a set $F \times A$, one can look for large subsets S on which T is 1-1. We will ask that S be large in a rather strong sense. For $S \subseteq F \times A$, we get a counting function, defined on X, given by

$$c_S(x) = \#\{g \in F; (g,x) \in S\}.$$

Of course

$$(c \times \mu)(S) = \int_A c_S(x)\, d\mu(x).$$

Set

$$c(S) = \min_{x \in A} c_S(x).$$

Definition 3.0.4. *We say that* $F \times A$ **maps an ε-quasi-tower** *if there exists a measurable subset* $S \subseteq F \times A$ *such that:*

(1) $T|_S$ *is 1-1; and*
(2) $c(S) \geq (1 - \varepsilon)\#F$.

The ε-quasi-tower itself is $T(F \times A) \subseteq X$. Notice that we may always assume $T(S) = T(F \times A)$. Note that if there exists an $S \subseteq F \times A$, such that T is 1-1 on S and $(c \times \mu)(S) > (1-\varepsilon^2)(c \times \mu)(F \times A)$, then there must exist an $A' \subseteq A$, with $\mu(A') > (1 - \varepsilon)\mu(A)$, such that $c_S(x) > (1 - \varepsilon)\#F$, for all $x \in A'$. Hence $F \times A'$ maps to an ε-quasi-tower.

We now state our version of the Ornstein–Weiss Rokhlin lemma. It is only a minor modification of Theorem 5, II.2 of [37].

Theorem 3.0.5 ([37]). *Suppose G is a discrete amenable group. For any $\varepsilon > 0$, there exist $\delta > 0$, $K \subseteq G$ and $N = N(\varepsilon)$ such that for any sequence $H_1, \ldots, H_N$ of (δ, K)-invariant subsets of G, and any free measure-preserving G-action $T = \{T_g\}_{g \in G}$, acting on $(X, \mathscr{B}, \mu)$, there exist sets $A_1, \ldots, A_N \in \mathscr{B}$ such that:*

(1) *each $H_i \times A_i$ maps to an ε-quasi-tower $\mathscr{R}_i$ in X;*
(2) *for $i \neq j$, $\mathscr{R}_i \cap \mathscr{R}_j = \emptyset$; and*
(3) $\mu(\cup_{i=1}^N \mathscr{R}_i) > 1 - \varepsilon$.

A collection of sets of the form $\{H_i \times A_i\}_{i=1}^{N}$ satisfying (1), (2) and (3), we call an **ε-Rokhlin tower**.

As we indicated earlier, Ornstein and Weiss prove this for a slightly different notion of ε-quasi-tower. We commented above that using ε^2 in their result gives the ε in ours.

Their statement differs slightly in another respect. They partition each A_i further into sets $A_{i,j}$ with T actually 1-1 on each $H_i \times A_{i,j}$. In fact, for any set $A \in \mathscr{B}$, and finite set $H \subseteq G$, A can be partitioned into a countable list of sets A_j with T 1-1 on each $H \times A_j$ simply because T acts freely. Hence this added structure is automatic.

Theorem 3.0.6 ([37]). *Given any finite partition P of X, one can select the sets A_i in Theorem 3.0.5 with*

$$A_i \perp \bigvee_{g \in H_i} T_{g^{-1}}(P).$$

Proof This result is Theorem 6, II.2 of [37]. Again, we have stated a slightly strengthened version. The important observation from Theorem 6, II.2, is that setting "ε"= ε/10, one obtains N and (δ, K) for Theorem 3.0.5. Then for (δ, K)-invariant sets $H_1, \ldots, H_N$, one is now at liberty to choose a finite partition Q, which we set to be

$$\bigvee_{g \in \bigcup_i H_i} T_{g^{-1}}(P).$$

Theorem 6, II.2 contains an extra parameter δ with $A_i \perp^{\delta} Q$. But as Ornstein and Weiss point out following the proof, by slightly shrinking the A_i, one obtains strict independence. □

We now describe the entropy of an ergodic, measure-preserving G-action. We use a name-counting approach, as described in [44]. Because we are considering only discrete amenable group actions, the entropy function we describe appears simpler than that described by Ornstein and Weiss in [37], where they consider names that take values in some compact metric space. Here we will count names that take values in some finite partition (or state space). In fact, though our approach appears to be different, the entropy function we describe is the same as that described in [37].

Definition 3.0.7. *A finite partition of a set X is a map P from X to some finite symbol space. We will write the symbol space as Σ_P which is simply* *the range of P. By the cardinality #P we simply mean the cardinality of* Σ_P.

Later we will consider partitions where the nature of the state space is important, but for considerations of entropy it is simply their cardinality that will come into play.

Nonetheless it is convenient to use symbol spaces as, for example, the partition $P \vee Q$ is then easily described as the tensor product of the two partitions.

We can topologize the Σ-valued partitions of a fixed space X with an L^1 metric $\|P_1, P_2\|_1 = \mu\{x; P_1(x) \neq P_2(x)\}$. This is usually referred to as the partition metric.

We can also topologize the partitions taking values in Σ with a distribution topology. This is actually a pseudo-metric topology. As we saw when we considered the distribution topology on rearrangements, this is a weak*-topology on measures.

To any finite partition P labeled by Σ we can consider the map to names.

$$\vec{P}(x) = \{P(T_g(x))\}_{g \in G} \in \Sigma^G.$$

As Σ^G is compact and metrizable the dual of the continuous functions is precisely the space of Borel measures. The shift action σ of G acts on names in Σ^G and $\vec{P}$ conjugates the action of T to that of σ. Hence $\vec{P}^*(\mu)$ is a σ-invariant Borel probability measure. Hence we can pseudo-topologize the "process" (that is to say, a G-action $(X, \mathcal{F}, \mu, T)$ and partition P (usually abbreviated to just (T, P)) with the weak*-topology on $\vec{P}^*(\mu)$. We give an explicit metric giving this topology as follows. For a Følner sequence $\{F_i\}$, for each i, consider the set of finite names $N_i = \{n : F_i \to \Sigma\}$. For each i and $n \in N_i$ define the cylinder set associated with such a name by

$$C(n) = \{\vec{n} \in \Sigma^G : \vec{n}(g) = n(g) \text{ for all } g \in F_i\}.$$

All such sets are clopen and so their characteristic functions are continuous. Furthermore, the finite linear combinations of such characteristic functions are uniformly dense in the space of continuous functions. Hence to say $m_i \to m$ weak* for Borel measures on $\{1, \ldots, N\}^G$ is equivalent to saying $m_i(C(n)) \to m(C(n))$ for all names $n \in N_i$ and all i.

For each i we can define a pseudometric on the Borel probability measures on Σ^G by

$$\|m_1, m_2\|^i = \frac{1}{2} \sum_{n \in N_i} |m_1(C(n)) - m_2(C(n))|.$$

It is an easy calculation to show that this is a pseudometric and is uniformly bounded by 1. (There are natural reasons for the coeffiecient 1/2 which we will not go into here.)

Notice that as F_i is nested and increasing, the values $\|m_1, m_2\|^i$ are non-decreasing in i.

Define a metric on the probability measures on Σ^G by

$$\|m_1, m_2\|_* = \sum_{i=1}^{\infty} \|m_1, m_1\|^i 2^{-i}.$$

Our earlier discussion makes it clear that this metric gives the weak*-topology on Borel measures. By pulling it back to

$$\|(T, P), (T', P')\|_* = \|(\vec{P})^*(\mu), (\vec{P}')^*(\mu')\|_*$$

we get an explicit metric giving the distribution topology on processes.

We now begin our description of entropy. Let $P : X \to \Sigma_P$ be any finite partition of X. As earlier we lift P to a map $\vec{P} : X \to \Sigma_P^G$ given by

$$(\vec{P}(x))(g) = P(T_g x), \text{ for all } g \in G.$$

We call $\vec{P}(x)$ the ***T*, *P*-name of *x***. For any finite set $F \subseteq G$, we may restrict $\vec{P}$ to F to get a map $\vec{P}_F : X \to \Sigma_P^F$. We say that $\vec{P}_F(x)$ is the ***T*, *P*, *F*-name of *x***.

Let $\{F_i\}$ be a nested Følner sequence with $\{F_i\} \nearrow G$ and $\mathrm{id} \in F_i$, for all i. For each fixed F_i, there are at most $\#P^{\#F_i}$ possible T, P, F_i-names covering X.

Let $\mu \in \mathcal{M}(X)$. Let $\varepsilon > 0$. Starting with the names of least μ-measure, remove as many as possible, in such a way that the μ-measure of the union of the remaining names is still greater than $1 - \varepsilon$. Let $S(T, P; F_i, \varepsilon)$ be the collection of remaining names. Let $N(T, P; F_i, \varepsilon) = \#S(T, P; F_i, \varepsilon)$. Notice that $N(T, P; F_i, \varepsilon) \leq \#P^{\#F_i}$.

Define

$$h(T, P; F_i, \varepsilon) = \frac{1}{\#F_i} \log_2 N(T, P; F_i, \varepsilon).$$

Certainly $h(T, P; F_i, \varepsilon) \leq \log_2(\#P)$. Define

$$h(T, P) = \lim_{\varepsilon \to 0} \liminf_{i \to \infty} h(T, P; F_i, \varepsilon).$$

Lemma 3.0.8. *For T ergodic, P a finite partition and for any ε' with $\varepsilon' < \varepsilon < 1$, we have*

$$\limsup_{i\to\infty} h(T,P;F_i,\varepsilon') \le \liminf_{i\to\infty} h(T,P;F_i,\varepsilon).$$

Proof Let $\varepsilon > 0$, with $\varepsilon < 1$. Let $\varepsilon' < \varepsilon$ and let $\eta < \varepsilon'$. In the following, for brevity, we use the notation $\{F_i\}$, even after passing to a subsequence. Select a subsequence of $\{F_i\}$ such that

$$\lim_{i\to\infty} h(T,P;F_i,\varepsilon) = \liminf_{i\to\infty} h(T,P;F_i,\varepsilon) = h(T,P;\varepsilon).$$

We may assume that this subsequence is sufficiently invariant. By Theorem 3.0.3, this means that these Følner sets may be used for quasi-tiling. To be more specific, use Theorem 3.0.3 to select K so that if $F_1,\dots,F_K$ is a sufficiently invariant sequence, then for any sufficiently invariant set F, F may be η-quasi-tiled by $F_1,\dots,F_K$.

Fix such sufficiently invariant $F_1,\dots,F_K$ so that for all $i = 1,\dots,K$,

$$|h(T,P;F_i,\varepsilon) - h(T,P;\varepsilon)| < \frac{\eta}{10}.$$

Fix F, sufficiently invariant to be η-quasi-tiled by $F_1,\dots,F_K$. Then for each $i = 1,\dots,K$, we have that

$$N(T,P;F_i,\varepsilon) \le 2^{(h(T,P;\varepsilon)+\frac{\eta}{10})\#F_i}.$$

There must exist centers $c_{i,j}$, $i = 1,\dots,K$, $j = 1,\dots,l(i)$, and subsets $F_{i,j} \subseteq F_i$ such that:

(1) $|F_{i,j}| \ge (1-\eta)\#F_i$, for $j = 1,\dots,l(i)$;
(2) the $F_{i,j}c_{i,j} \subseteq F$ are disjoint; and
(3) $\#(\bigcup_{i,j} F_{i,j}c_{i,j}) \ge (1-\eta)\#F$.

We want to count the "good" names across F. Specifically, we want to estimate $N(T,P;F,\eta)$.

There exists a set $D \subseteq X$ with $\mu(D) > 1-\eta/\varepsilon$ such that for $x \in D$,

$$\sum_{i,j} \#F_{i,j} 1_{S(T,P;F_i,\eta)}(T_{c_{i,j}}x) > (1-\varepsilon)\sum_{i,j} \#F_{i,j}.$$

We will consider names of points x in D. The number of such T,P,F-names of points in D is bounded by the product of three terms, as follows:

The first term is the number of ways the sets $F_1,\dots,F_K$ could arise, which is bounded above by the number of subsets of size at most $\eta\#F$

in a set of size $\#F$. Standard Stirling's formula estimates show that this term is bounded above by

$$2^{(H(\eta)\#F+\frac{\log_2 \#F}{2})},$$

where, as usual,

$$H(\alpha) = -\alpha \log_2 \alpha - (1-\alpha)\log_2(1-\alpha).$$

The second term is the product over all i, j of the number of possible good names across each $F_{i,j}c_{i,j}$. This is bounded above by $2^{(h(T,P;\varepsilon)+\frac{\eta}{10})\#F}$.

Finally, the third term is the number of possible names outside the tiling, which is bounded above by

$$\#P^{\eta\#F}.$$

Putting this all together, the number of T, P, F-names covering all but η of X is

$$N(T,P;F,\eta) \leq 2^r$$

where

$$r \leq \#F[H(\eta) + \frac{\log_2(\#F)}{2\#F} + \eta \log_2 \#P + \frac{\eta}{10} + h(T,P;\varepsilon)].$$

This holds for all sufficiently invariant F and thus

$$h(T,P;F,\eta) \leq H(\eta) + \eta \log_2 \#P + \frac{\eta}{10} + h(T,P;\varepsilon).$$

Letting $\eta \to 0$, since $\eta \leq \varepsilon'$, we have that

$$\limsup_{i\to\infty} h(T,P;F_i,\varepsilon') \leq h(T,P;\varepsilon),$$

which completes the proof. □

Note that the preceding argument shows that the definition of entropy is independent of the choice of Følner sequence.

Define

$$h(T) = \sup\{h(T,P); P \in \mathscr{P}\}.$$

The Ornstein–Weiss version of a Shannon–McMillan type theorem, in our situation, reads as follows ([37], Section II.4, Theorem 5).

Theorem 3.0.9. *Suppose $(X, \mathscr{B}, \mu, T)$ is an ergodic G-action (G a discrete amenable group). Suppose P is any finite partition of X. For any $\varepsilon > 0$, if $F \subseteq G$ is sufficiently invariant, there exist T, P, F-names $\bar{P}_1, \bar{P}_2, \ldots, \bar{P}_k$ such that:*

(1) $\mu(\cup_{i=1}^{k}\bar{P}_i) > 1 - \varepsilon$;
(2) $k \leq 2^{(h(T,P)+\varepsilon)\#F}$; *and*
(3) *for all* i, $\mu(\bar{P}_i) \leq 2^{-(h(T,P)-\varepsilon)\#F}$.

It will be helpful to restate this in a slightly different form:

Corollary 3.0.10. *Suppose* $(X, \mathscr{B}, \mu, T)$ *is an ergodic G-action (G a discrete amenable group). Suppose* P *is any finite partition of* X. *For any* $\varepsilon > 0$, *if* $F \subseteq G$ *is sufficiently invariant, there exist* T, P, F*-names* $\bar{P}_1, \bar{P}_2, \ldots, \bar{P}_k$ *such that:*

(1) $\mu(\cup_{i=1}^{k}\bar{P}_i) > 1 - \varepsilon$; *and*
(2) *for all* i, $\mu(\bar{P}_i) = 2^{-(h(T,P)\pm\varepsilon)\#F}$.

Proof Just notice that for the $\bar{P}_i$ of Theorem 3.0.9 we can calculate that the measure of the union of all $\bar{P}_i$ whose measure is less than $2^{-(h(T,P)+2\varepsilon)\#F}$ is at most $2^{-\varepsilon\#F}$. For F to be very invariant it must be large, in particular large enough to make $2^{-\varepsilon\#F} < \varepsilon$. This says that we can obtain the result for 2ε just by taking the list of $\bar{P}_i$ from Theorem 3.0.9 for ε and deleting the elements that are too small. □

From this, we get the following useful corollary.

Corollary 3.0.11. *Suppose* $(X, \mathscr{B}, \mu, T)$ *is an ergodic G-action. Suppose* P *is any finite partition of* X. *Let* $\varepsilon < 1$. *Then*

$$h(T,P) = \lim_{i\to\infty} h(T,P;F_i,\varepsilon).$$

Proof Theorem 3.0.9 tells us that for all $\varepsilon > 0$, there exists I such that for all $i \geq I$ and $\delta < 1$, $N(T,P;F_i,\delta)$ lies within

$$2^{(h(T,P)\pm\varepsilon)\#F_i}$$

so that

$$h(T,P;F_i,\delta)$$

lies within $h(T,P) \pm \varepsilon$. □

Note that the corollary implies $\lim_{\varepsilon\to 0}\lim_{i\to\infty} h(T,P;F_i,\varepsilon) = h(T,P)$.

Notice that the entropy of a process $h(T,P)$ is actually a function of the image measure $\vec{P}^*(\mu)$ as it only depends on the measures of finite names. Thus we can write $h(\tilde{\mu})$ instead of $h(T,P)$ where $\tilde{\mu} = \vec{P}^*(\mu)$. For our subsequent work, we will need to know that not only entropy, but conditional entropy, as a function of this measure is upper semi-continuous in the weak*-topology. This is a very well-known fact. In particular Weiss has shown that it is precisely the amenable group actions for which it is true. As the proof is not available elsewhere we include the counting argument that leads to it. As with the Shannon–McMillan theorem in this context, it is a direct consequence of the Ornstein–Weiss quasi-tiling theorem, through its corollaries, the mean ergodic theorem and the Shannon–McMillan theorem.

Definition 3.0.12. *For $P : X \to \Sigma_P$ and $Q : X \to \Sigma_Q$ two finite partitions of a free and ergodic G-action $(X, \mathcal{F}, \mu, T)$ we define the conditional entropy as follows.*

For any finite set $F \subseteq G$, finite partition R and $x \in X$ let

$$C_F^R(x) \in \vee_{g \in F} T_{g^{-1}}(R)$$

be the cylinder containing x.

Now let

$$N(T, P|Q; F, \varepsilon)(x)$$

be the minimum number of elements of $\bigvee_{g \in F} T_{g^{-1}}(P \vee Q)$ it takes to cover all but a fraction ε in measure of the cylinder $C_F^Q(x)$.

Note: If Q is the trivial partition this is $N(T, P; F, \varepsilon)$.

Set

$$h(T, P|Q; F, \varepsilon)(x) = \frac{1}{\#F} \log_2(N(T, P|Q; F, \varepsilon)(x)).$$

Lemma 3.0.13. *The functions $h(T, P|Q; F, \varepsilon)$ are all bounded by*

$$\log_2(\#\Sigma_P \times \#\Sigma_Q).$$

For $\{F_i\}$ a Følner sequence and any $\varepsilon < 1$,

$$h(T, P|Q; F_i, \varepsilon)(x) \underset{i}{\to} h(T, P \vee Q) - h(T, Q)$$

in distribution and hence in L_1.

Proof From Corollary 3.0.10 for any $\varepsilon > \bar{\varepsilon} > 0$, if F_i is sufficiently invariant, then for all but $\bar{\varepsilon}^2$ in measure of the $x \in X$,

$$\mu(C_{F_i}^Q(x)) = 2^{(h(T,Q) \pm \bar{\varepsilon}^2)\#F_i}$$

and

$$\mu(C_{F_i}^{P \vee Q}(x)) = 2^{-(h(T,P \vee Q) \pm \bar{\varepsilon}^2)\#F_i}.$$

We call these the "good" Q and $P \vee Q$ atoms, respectively.

Hence for all but $2\bar{\varepsilon}$ of the $x \in X$, $C^Q_{F_i}$ is a good Q atom and is all but a fraction $\bar{\varepsilon}$ in measure covered by good $P \vee Q$ atoms. Thus any subset of all but a fraction ε of $C^Q_{F_i}(x)$ must contain at least

$$2^{(h(T,P\vee Q)-h(T,Q)-2\bar{\varepsilon})\#F_i}(1-\varepsilon-\bar{\varepsilon})$$

elements of $\bigvee_{g\in F_i} T_{g^{-1}}(P \vee Q)$ (the "good" atoms still in the set).

Furthermore, as $\bar{\varepsilon} < \varepsilon$, all but a fraction ε is indeed covered by fewer than

$$2^{(h(T,P\vee Q)-h(T,Q)+2\bar{\varepsilon})\#F_i}$$

atoms of $\bigvee_{g\in F_i} T_{g^{-1}}(P \vee Q)$ (the good atoms).

Hence once F_i is sufficiently invariant, for all but $\bar{\varepsilon}$ of the $x \in X$,

$$\begin{aligned} h(T,P\vee Q) - h(T,Q) - 2\bar{\varepsilon} \le\, & h(T,P|Q;F_i,\varepsilon)(x) \\ \le\, & h(T,P\vee Q) - h(T,Q) + 2\bar{\varepsilon} \\ & + \log_2(1-\varepsilon-\bar{\varepsilon})/\#F_i. \end{aligned}$$

The result follows by choosing $\bar{\varepsilon}$ small enough that $1-\varepsilon-\bar{\varepsilon} > 0$ and letting $F_i \nearrow G$. □

Definition 3.0.14. *We define the conditional entropy $h(T,P|Q)$, where P and Q are finite partitions, by*

$$h(T,P|Q) = h(T,P\vee Q) - h(T,Q).$$

Theorem 3.0.15. *For any $\varepsilon > 0$ and bound B there are values $\delta > 0$ and N and a finite set $K \in G$ so that if $H_1, H_2, \dots H_N$ are a δ, K-invariant sequence of finite sets in G and T is a free and ergodic G-action on the probability space $(X, \mathscr{F}, \mu)$ and P and Q are finite partitions of X with both $\#\Sigma_P$ and $\#\Sigma_Q < B$ satisfying*

$$\|h(T,P|Q;H_i,\delta)\|_1 \le E$$

for all $i = 1, \dots, N$ then

$$h(T,P|Q) \le E + \varepsilon.$$

Proof Fix a value ε_1, which we will set later. Choose $\delta > 0$, $K \subset G$ and N so that for any δ, K-invariant sequence $H_1, \dots, H_N$ in any free and ergodic action $(X, \mathscr{F}, \mu, T)$ and finite partitions P and Q we can constuct an ε_1-Rokhlin tower (Theorem 3.0.6) $\{H_i \times A_I\}$ with

$$A_i \perp \vee_{g \in H_i} T_{g^{-1}}(P \vee Q).$$

Also require that for all i,

$$\frac{1}{\#H_i(1-\varepsilon_1)} < \varepsilon_1$$

forcing

$$\mu(\cup_i A_i) < \varepsilon_1.$$

For each $i = 1, \ldots, N$ let G_i consist of those x for which $C_i^{P \vee Q}(x)$ is one of the $N(T, P|Q; H_i, \delta)$ atoms we know can cover all but a fraction δ of $C_i^Q(x)$. As G_i is $\bigvee_{g \in H_i} T_{g^{-1}}(P \vee Q)$-measurable,

$$\mu(A_i \cap G_i) = \mu(A_i)\mu(G_i).$$

Define a function

$$N(x) = \begin{cases} \log_2(N(T, P|Q; H_i, \delta))(x) & \text{if } x \in A_i \cap G_i \text{ and} \\ 0 & \text{otherwise.} \end{cases}$$

Notice that

$$\begin{aligned} \int N \, d\mu &= \sum_{i=1}^{N} \mu(A_i) \int \log_2(N(T, P|Q; H_i, \delta)) \, d\mu \\ &= \sum_{i=1}^{N} \mu(A_i) \# H_i \left(\int h(T, P|Q; H_i, \delta) \, d\mu \right) \\ &\le E \sum_{i=1}^{N} \mu(A_i) \# H_i \le E(1 + \varepsilon_1). \end{aligned}$$

The mean ergodic theorem (theorem 3, II.1 of [37]) now tells us that for all F sufficiently invariant, for all but ε_1 of the $x \in X$ the following three conditions hold. First,

$$\frac{1}{\#F} \sum_{g \in F} N(T_g(x)) \le E(1 + \varepsilon_1) + \varepsilon_1$$

from the above calculation. Applying it to two characteristic functions we also ask that second,

$$\#\{g \in F : T_g(x) \text{ is in the tower image } \} \ge (1 - 2\varepsilon_1)\#F$$

and third,

$$\#\{g \in F : T_g(x) \in \cup_i A_i\} \le \mu(\cup_i A_i)(1 + \varepsilon_1)\#F \le \varepsilon_1(1 + \varepsilon_1)\#F.$$

Call the set of points in x that satisfy all three of the above conditions $D(F)$.

We also ask that F be sufficiently invariant that for all $i = 1, \ldots, N$,

$$\frac{\#(H_i F \triangle F)}{\#F} < \frac{\varepsilon_1}{N}.$$

This now implies that for all but $\sqrt{\varepsilon_1}$ of the $x \in X$,

$$\mu(C_F^Q(x) \cap D(F)) \geq \mu(C_F^Q(x))(1 - \sqrt{\varepsilon_1}).$$

We now estimate the number of atoms of $\bigvee_{g \in F} T_{g^{-1}}(P \vee Q)$ it takes to cover $C_F^Q(x) \cap D(F)$.

For $x' \in C_F^Q(x) \cap D(F)$ consider the set of values $g \in F$ such that:

(1) $T_g(x) \in \cup_i A_i \cap G_i$ (i.e. $T_g(x)$ is the center of a "tile"); and

(2) $H_i g \subseteq F$ for all i (i.e. the entire tile centered at g lies in F).

The number of ways to select the set of values g to be such centers is

$$\leq \binom{\#F}{\varepsilon_1(1+\varepsilon_1)\#F} \underset{\text{def}}{=} K.$$

Having split according to where the centers lie, we now further subdivide according to how each center is assigned to one of the potential tiles H_i, that is to say according to which $A_i \cap G_i$ actually contains the center g. The number of ways to do this is bounded by

$$N^{\varepsilon_1(1+\varepsilon_1)\#F}.$$

Fix a collection of choices for centers, and for the tile H_i each is the center of. Let them be

$$c_{1,i}, c_{2,i}, \ldots, c_{\ell(i),i} \in A_i \cap G_i,$$

and let $x' \in C_F^Q \cap D(F)$ be a point for which these list the actual g's satisfying (1) and (2) above.

The set $\cup_{i,j} H_i c_{i,j}$ covers all but a fraction $3\varepsilon_1$ of F as it contains all $g \in F$ with $T_g(x')$ in the tower image, except for those too near the boundary of F to put the full slice into F.

We also know that if $g \in F$ and $T_g(x') \in A_i \cap G_i$ and $H_i g \subseteq F$ then

$g = c_{i,j}$ for some j. Hence

$$
\begin{aligned}
E(1+\varepsilon_1)+\varepsilon_1 &\geq \frac{1}{\#F}\sum_{g\in F} N(T_g(x)) \\
&\geq \frac{1}{\#F}\log_2\left(\Pi_{i,j} N(T, P|Q; H_i, \delta)(T_{c_{i,j}}(x'))\right) \\
&= \frac{1}{\#F}\log_2\left(\Pi_{i,j}(\#\text{of choices for the } P\vee Q \text{ name on } H_i c_{i,j})\right) \\
&\geq \frac{1}{\#F}\log_2\left(\#\text{of choices for the } P\vee Q \text{ name on } \cup_{i,j} H_i c_{i,j}\right).
\end{aligned}
$$

The number of possible $P \vee Q$ names that might occur on the remainder of F outside of $\cup_{i,j} H_i c_{i,j}$ is certainly at most

$$\left(\#\Sigma_P \times \#\Sigma_Q\right)^{3\varepsilon_1 \#F} \leq B^{6\varepsilon_1 \#F}.$$

Combining these estimates we conclude that for all but $\sqrt{\varepsilon_1}$ of the $x \in X$,

$$
\begin{aligned}
\frac{1}{\#F}\log_2(N(T, P|Q; F, \sqrt{\varepsilon_1})(x)) \leq \frac{\log_2(K)}{\#F} &+ \varepsilon_1(1+\varepsilon_1)\log_2(N) \\
&+ 6\varepsilon_1 \log_2(B) + E(1+\varepsilon_1) + \varepsilon_1.
\end{aligned}
$$

Standard Stirling's formula estimates tell us that if ε_1 is small enough then $\log_2(K)/\#F < \varepsilon/3$. If ε_1 is small enough, depending only on N, B and E, all the other error terms in this estimate can be forced to contribute less than $\varepsilon/3$ as well.

We conclude that once F is sufficiently invariant, then for all but $\sqrt{\varepsilon_1}$ of the $x \in X$,

$$h(T, P|Q; F, \sqrt{\varepsilon_1})(x) \leq E + 2\varepsilon/3.$$

The previous Lemma (3.0.13) now guarantees the conclusion. □

Definition 3.0.16. *Notice that as with entropy the calculation of conditional entropy depends only on the measure*

$$\tilde{\mu} = (\overrightarrow{P \vee Q})^*(\mu)$$

on $(\Sigma_P \times \Sigma_Q)^G$. *As a function of such* σ*-invariant and ergodic measures* m *we write this conditional entropy as*

$$h(\tilde{\mu}, \Sigma_P|\Sigma_Q).$$

Theorem 3.0.17. *For any two finite symbol spaces Σ_P and Σ_Q the function $h(\nu, \Sigma_P|\Sigma_Q)$ on the σ-invariant and ergodic measures on $(\Sigma_P \times \Sigma_Q)^G$ is upper semi-continuous.*

Proof To begin, as we can always take the direct product of a non-free ergodic G-action with a Bernoulli action, getting a free and ergodic G-action, Theorem 3.0.15 holds for any ergodic action, whether free or not. Next, fix a σ-invariant and ergodic measure ν. Suppose that

$$h(\nu, P|Q) = E$$

and by Lemma 3.0.13 we know that for any $\varepsilon < 1$, for any Følner sequence F_i,

$$h(\nu, P|Q; F_i, \varepsilon) \underset{i}{\rightarrow} E$$

in $L^1(\nu)$. Let $\varepsilon > 0$ be chosen, and let B bound the cardinalities of both P and Q. Using $\varepsilon/3$ as "ε" in Theorem 3.0.3 we obtain a δ, K and N. Among the terms of the Følner sequence $\{F_i\}$ we then will be able to pick a list $H_1, H_2, \ldots, H_N$ that forms a δ, K-invariant sequence and for which

$$h(\nu, P|Q; H_i, \delta) \leq E + \varepsilon/3.$$

There are only finitely many clopen sets $C_{H_i}^{P \vee Q}$ and so once ν' is sufficiently close to ν weak* we will have

$$h(\nu, P|Q; H_i, \delta) \leq E + 2\varepsilon/3.$$

Theorem 3.0.15 now implies that if ν' is this close weak* to ν, then

$$h(\nu, P|Q) \leq E + \varepsilon.$$

This completes the result.

We can apply this result to rearrangements with a simple observation. Suppose $(X, \mathscr{F}, \mu, T^\alpha)$ is a free and ergodic action of G, and P and Q are two finite partitions of X. For any arrangement β we will have a corresponding map to names

$$(P \overrightarrow{\vee} Q)_\beta(x) = \{P(T_g^\beta(x)), Q(T_g^\beta(x))\}_{g \in G}$$

and hence a measure

$$\tilde{\mu}(\beta, P \vee Q) = (P \overrightarrow{\vee} Q)_\beta^*(\mu).$$

It is little more than an observation that the map $\beta \to \tilde{\mu}(\beta, P \vee Q)$ is continuous as a map from the $\|\cdot,\cdot\|_1$-topology to the weak*-topology. With a bit more thought one is easily convinced that it is uniformly continuous in the metrics we have established. (This latter remark will not be critical for us.)

Corollary 3.0.18. *For P and Q two finite partitions of the free and ergodic G-action $(X, \mathscr{F}, \mu, T)$, the conditional entropy $h(T^{\alpha}, P|Q)$ is upper semi-continuous as a function of α in the $\|\cdot,\cdot\|_1$-topology. That is to say if $\alpha_i \to \alpha$ in L^1 then*

$$\limsup_{i \to \infty} h(T^{\alpha_i}, P|Q) \le h(T^{\alpha}, P|Q).$$

4
Copying Lemmas

Copying lemmas play a pivotal role in Dye's Theorem [10], Vershik's Lacunary Isomorphism Theorem [58] and most significantly in Ornstein's Isomorphism Theorem for Bernoulli shifts [32]. The Burton–Rothstein version of this last result puts them in an even more central role as the rest of the argument becomes soft analysis. We are generalizing from this Burton–Rothstein perspective of course, making the core of our equivalence theorem rest on category. Copying lemmas will play a pivotal role for us in two contexts. First in the equivalence theorem they will play the same role as always, allowing one to copy partitions, and in our case full-group elements, from a joining of two systems into one of the two. We will also use a copying lemma as a basic tool in our development of m-entropy to show that sizes are either "entropy-preserving" or "entropy-free".

These two applications have one fact in common. As indicated above one must copy not only partitions but full-group elements. Hence we will have to investigate how one does this. These copying lemmas will also have one very real difference. For use in the equivalence theorem one will want the copied process to have as much entropy as one can hope for, but for use in the entropy theory, one will want the copy to kill as much entropy as possible. Hence we really must give two copying lemmas. They will be parallel in structure, but in assigning names in one case we will try to make the assignment as close to 1-1 as possible, and in the second, as far from 1-1 as possible.

We have tried to make our copying lemmas "modern", avoiding the unnecessary use of Hall's Marriage lemma and making no minimum entropy-bound assumption on the image system. We do more here than is essential for our work, hoping to provide a tool-box of copying lemmas. The basic technical lemmas are Theorems 4.0.5 and 4.0.13. We establish various corollaries of these that state explicitly the precise technical versions we will use.

Let $R = \{H_i \times A_i\}_{i=1}^N$ be an ε-Rokhlin tower for T (see Theorems 3.0.5 and 3.0.6). For $x \in A_i$ we set

$$S(x) = \{x' \in T(H_i \times \{x\}) : x' \text{ has a unique preimage in } H_i \times \{A_i\}\}.$$

These are the **slices** through the tower and are the atoms of a measurable partition of the **tower image**, given by

$$\{x' \in T(\bigcup_i(H_i \times A_i)) : x' \text{ has a unique preimage in } \bigcup_i (H_i \times A_i)\}.$$

We set

$$g(x) = \{g \in H_i : T_g(x) \in S(x)\},$$

the slice viewed as a subset of $H_i \subseteq G$.

Suppose Σ is some finite labeling set and $n_i : A_i \to \Sigma^{H_i}$ assigns to each $x \in A_i$ a **Σ, H_i-name**, that is to say, an element in Σ^{H_i}. We want to describe the notion of **painting** the names n_i onto the tower $\{H_i \times A_i\}_{i=1}^N$.

For any $x' \in S(x)$ and $x \in A_i$ there exists $g \in H_i$ such that $x' = T_g(x)$. Set $P(x') = n_i(x)(g)$, thus defining a map

$$P : \text{tower image} \to \Sigma.$$

To complete its definition we must extend P outside the tower image. We will discribe how to do this, depending on the circumstances. If no explicit description is given, then any extension will do.

As described twice earlier, first in Section 2.1 when discussing rearrangements, and later in Chapter 3 when developing entropy, for a choice of symbol space Σ, finite or countably infinite, we can place two measures of closeness (one a metric, the other a pseudometric) on Σ-valued partitions. These generalize what we have already described as the L^1 and distribution topologies on rearrangements and on finite partitions. Our description here is parallel to that given earlier.

If $P_1, P_2 : X \to \Sigma$ then

$$\begin{aligned}\|P_1, P_2\|_1 &= \frac{1}{2}\sum_{n_0 \in \Sigma} \mu(P_1^{-1}(n_0) \triangle P_2^{-1}(n_0)) = \mu(\{x : P_1(x) \neq P_2(x)\}) \\ &= \int d(P_1(x), P_2(x))\, d\mu\end{aligned}$$

where d is the discrete 0,1 valued metric on Σ. For $P : X \to \Sigma$ any partition, we can define a map $\vec{P} : x \to \Sigma^G$ given by

$$\vec{P}(x) = \{P(T_g(x))\}_{g \in G}.$$

This is just the map taking x to its **name**. If Σ is finite, then Σ^G will be compact in the product topology. If Σ is countably infinite, adjoining a single point at infinity, it becomes compact, and again we can regard Σ^G as compact. Hence the space of probability measures on Σ^G is a compact and convex metric space in the weak*-topology.

The distribution pseudotopology on Σ-valued processes is this weak*-topology on the Borel measures $\vec{P}^*(\mu)$. It will be useful to have in mind a way to verify that two processes are in fact close in distribution. We lift to this context the distribution metric described earlier for both rearrangements and finite partitions. For F any finite subset of G, the map $P_F : x \to \{P(T_g(x))\}_{g \in F}$ maps $X \to \Sigma^F$ which is an at most countable set. Thus $P_F^*(\mu)$ can be regarded as simply a vector of masses on Σ^F, and one can calculate the ℓ_1 distance between these vectors,

$$|(T_1, P_1), (T_2, P_2)|_F = \|P_{1,F}^*(\mu_1), P_{2,F}^*(\mu_2)\|_1.$$

For F_i some increasing sequence of finite sets, exhausting G (for example a Følner sequence), we can define

$$\|(T_1, P_1), (T_2, P_2)\|_* = \sum_{i=1}^{\infty} \frac{1}{2^i} |(T_1, P_1), (T_2, P_2)|_{F_i}.$$

Notice that as $F_i \subseteq F_{i+1}$,

$$|(T_1, P_1), (T_2, P_2)|_{F_i} \leq |(T_1, P_1), (T_2, P_2)|_{F_{i+1}}.$$

Thus

$$\|(T_1, P_1), (T_2, P_2)\|_* \leq |(T_1, P_1), T_2, P_2)|_{F_i} + 1/2^i \qquad \text{for all } i.$$

We now proceed to demonstrate our copying lemmas. It will be useful to restrict our attention to finite partitions.

If P is a finite partition, let Σ_P be its **range**, that is to say, its labeling set.

For a rearrangement (α, ϕ) we have already associated a countable partition

$$g_{(\alpha,\phi)}(x) = \alpha(x, \phi(x)) = f_x^{\alpha,\phi}(\mathrm{id}) \in G.$$

Notice that $f_x^{\alpha,\phi}$ is precisely the T^α, $g_{(\alpha,\phi)}$-name of the point x. We say a rearrangement is **bounded** if the partition $g_{(\alpha,\phi)}$ is finite, which is equivalent to saying the values $\alpha(x, \phi(x))$ lie in some finite subset of G.

Definition 4.0.1. *Given a tower $\{H_i \times A_i\}_{i=1}^N$, we say a rearrangement (α, ϕ)* **respects the tower** *if for any $x \in A_i, x' \in S(x)$, $\phi(x')$ is also in $S(x)$, and for any other x', outside the tower image, $\phi(x') = x'$.*

A rearrangement (α, ϕ) that respects a tower $\{H_i \times A_i\}_{i=1}^N$ will be bounded, and hence $g_{(\alpha,\phi)}$ will be a finite partition. Furthermore any ϕ which respects a tower will be of finite order as its cycles are at most slices through a tower.

Theorem 4.0.2. *For any rearrangement (α, ϕ) and $\varepsilon > 0$, there exists $K \subseteq G$ so that if $H_1, \ldots, H_N$ are $(\frac{\varepsilon}{4}, K)$-invariant and if $\{H_i \times A_i\}_{i=1}^N$ is an $\frac{\varepsilon}{4}$-Rochlin tower for T^α, then there exists $\phi' \in FG(\mathcal{O})$ so that*

(1) *(α, ϕ') respects the tower $\{H_i \times A_i\}_{i=1}^N$, and*
(2) $\|g_{(\alpha,\phi)}, g_{(\alpha,\phi')}\|_1 < \varepsilon$.

In particular the rearrangement (α, ϕ') is bounded and ϕ' is of finite order.

Proof Choose K finite so that $\mu(\{x : g_{(\alpha,\phi)}(x) \in K\}) > 1 - \frac{\varepsilon}{4}$. Suppose $H_1, \ldots, H_N$ are $(\frac{\varepsilon}{4}, K)$-invariant and $\{H_i \times A_i\}_{i=1}^N$ is an $\frac{\varepsilon}{4}$-Rokhlin tower for T^α.

We identify a set of **good points** C in each slice of the tower. For $x \in A_i$, $x' \in S(x)$, we say that $x' \in C$ if also $\phi(x') \in S(x)$. Notice that for $x \in A_i$,

$$
\begin{aligned}
S(x) \setminus C \subseteq \{x' :\ & \text{either} \\
& \text{(a) } \alpha(x', \phi(x')) \notin K \text{ or} \\
& \text{(b) } \alpha(x', \phi(x')) \in K, x' = T_g^\alpha(x) \text{ for some } g \in H_i, \\
& \qquad \text{but } \alpha(x', \phi(x'))g \notin H_i, \text{ or} \\
& \text{(c) } \phi(x') \in T(H_i \times A_i) \text{ but } \phi(x') \text{ doesn't have} \\
& \qquad \text{a unique preimage in } H_i \times A_i\}.
\end{aligned}
$$

Hence we compute that

$$\begin{aligned}
\mu(C) &\geq \mu\Big(\bigcup_{x\in\cup A_i} S(x)\Big) - \mu\Big(\bigcup_{x\in\cup A_i} S(x)\setminus C\Big)\\
&\geq \left(1-\frac{\varepsilon}{4}\right) - \mu\{x' : \text{(a) holds}\} - \mu\{x' : \text{(b) holds}\}\\
&\qquad - \mu\{x' : \text{(c) holds}\}\\
&\geq \left(1-\frac{\varepsilon}{4}\right) - \frac{\varepsilon}{4} - \sum_i \mu\Big\{\bigcup_{k\in K} T_k^{\alpha}(T(H_i\times A_i))\triangle T(H_i\times A_i)\Big\}\\
&\qquad - \mu\{\phi(x') : \text{(c) holds}\}\\
&\geq \left(1-\frac{\varepsilon}{4}\right) - \frac{\varepsilon}{4} - \sum_i \left(1-\frac{\varepsilon}{4}\right)\mu(A_i) - \frac{\varepsilon}{4}\\
&\geq 1-\varepsilon.
\end{aligned}$$

For $x' \in C$ define $\phi'(x') = \phi(x')$. Then for each slice $S(x)$ we have defined $\phi' : C \cap S(x) \to S(x)$. Furthermore

$$\begin{aligned}
&\#\{x' \in S(x) : \phi' \text{ is not yet defined at } x'\}\\
&\quad = \#\{x' \in S(x) : x' \text{ is not yet in the range of } \phi'\}.
\end{aligned}$$

Next, extend ϕ' to $\bigcup_i T(H_i \times A_i) \setminus C$ so that for each x, ϕ' is a 1-1 map from

$$\{x' \in S(x) : \phi' \text{ not yet defined at } x'\}$$

onto

$$\{x' \in S(x) : x' \text{ not yet in range of } \phi'\}.$$

This can be done measurably, since these two sets, as functions of x, were measurably chosen.

Finally, outside of $\cup_{i=1}^N T(H_i \times A_i)$, define $\phi' =$ identity.

This constructs $\phi' \in FG(\mathcal{O})$ with $\phi'|_C = \phi$, so that

$$\|g_{(\alpha,\phi)}, g_{(\alpha,\phi')}\|_1 < \varepsilon. \qquad \square$$

The following result will fill the gap for us between copying partitions and copying full-group elements.

Theorem 4.0.3. *Suppose* (α, ϕ) *is a rearrangement of the free and ergodic* G*-action* $(X, \mathcal{F}, \mu, T^{\alpha})$. *Given any* $\varepsilon > 0$ *there exists* $\delta > 0$ *(depending on* (α, ϕ) *as well) so that if* $P' : X_1 \to G$ *is a partition of the free and ergodic* G*-action* $(X_1, \mathcal{F}_1, \mu_1, T_1^{\alpha_1})$ *satisfying:*

$$\|(T^{\alpha}, g_{(\alpha,\phi)}), (T_1^{\alpha_1}, P')\|_* < \delta,$$

then there exists ϕ' in the full-group of $T_1^{\alpha_1}$ with

$$\|P', g_{(\alpha_1,\phi')}\|_1 < \varepsilon.$$

Notice that we could immediately also obtain:

$$\|(\alpha,\phi),(\alpha_1,\phi')\|_* < \varepsilon$$

as this will be implied if the "ε" and "δ" in Theorem 4.0.3 are small enough.

Proof As P' maps $X_1 \to G$, we can construct an associated element $f_{x_1} \in G^G$ by

$$f_{x_1}(g) = P'(T_{1,g}^{\alpha_1}).$$

The only issue is that $H(f_{x_1})$ might not be a bijection of G a.e. If it were then $\phi'(x_1) = T_{1,P'(x_1)}^{\alpha_1}(x_1)$ would already be the full-group element we are after.

Also note that if we construct ϕ' with

$$\mu(\{x_1 : \phi'(x_1) \neq P'(x_1)\}) < \varepsilon/2$$

then we will have

$$\|P', g_{(\alpha_1,\phi_1)}\|_1 < \varepsilon.$$

To begin, select $K \subseteq G$ so that

$$A = \{x : \alpha(x,\phi(x)) = f_x^{\alpha,\phi}(\mathrm{id}) \in K\}$$

has

$$\mu(A) > 1 - \varepsilon/4.$$

Select $K_1 \subseteq G$ with id $\in K_1$ so that for any $g \notin K_1$, id $\notin KgK$ (i.e. $K^{-1}K^{-1} \subseteq K_1$).

Now choose $\delta > 0$ so small that if

$$\|(T^\alpha, g_{(\alpha,\phi)}),(T_1^{\alpha_1},P')\|_* < \delta$$

then:

(i) Setting $A_1 = \{x_1 \in X_1 : P'(x_1) \in K\}$, we will still have

$$\mu_1(A_1) > 1 - \varepsilon/4$$

and:

(ii) Setting $G = \{x_1 \in X_1 :$ for **some** $x \in X$, $f_x^{\alpha,\phi}|_{K_1} = f_{x_1}|_{K_1}\}$ we will have $\mu_1(G) > 1 - \varepsilon/4$.

Set $G_1 = A_1 \cap G$ and $\mu_1(G_1) > 1 - \varepsilon/2$. We claim that the map

$$x_1 \to T^{\alpha_1}_{1,P'(x_1)}(x_i)$$

is 1-1 on G_1. To see this, suppose for x_1 and $x_1' = T^{\alpha_1}_{1,g}(x_1) \in G_1$ we have

$$T^{\alpha_1}_{1,P'(x_1)}(x_1) = T^{\alpha_1}_{1,P'(x_1')}(x_1').$$

This translates directly to the identity

$$x_1 = T^{\alpha_1}_{1,H(f_{x_1})(g)}(x_1),$$

which is to say

$$H(f_{x_1})(g) = f_{x_1}(g)gf_{x_1}(\mathrm{id})^{-1} = \mathrm{id}.$$

As both x_1 and x_1' are in G_1, both $f_{x_1}(\mathrm{id})$ and $f_{x_1}(g)$ are in K and we conclude that g must lie in K_1. But now since for some $x \in X$ we have

$$f_x^{\alpha,\phi}|_{K_1} = f_{x_1}|_{K_1},$$

we get that

$$f_{x_1}(g)gf_{x_1}(\mathrm{id})^{-1} = h_x^{\alpha,\phi}(g) = \mathrm{id}.$$

But for all x, $h_x^{\alpha,\phi}$ is a bijection of G, and so $g = \mathrm{id}$.

On the set G_1 set $\phi'(x_1) = T^{\alpha_1}_{1,P'(x_1)}(x_1)$. This is a 1-1 map, as we just saw, and consists piecewise of elements of $T_1^{\alpha_1}$. Hence it must be measure preserving. This means that the two sets G_1^c and $\phi'(G_1)^c$ have the same measure, which is in fact less than $\varepsilon/2$. It is a standard fact, and also a rather simple exercise using the Ornstein–Weiss Rokhlin lemma and mean ergodic theorem, that in a free and ergodic system like the $T_1^{\alpha_1}$ system there will always be full-group elements taking any one set to any other, as long as the two sets have the same measure. This means that ϕ' can be extended to a full-group element that takes G_1^c to $\phi'(G_1)^c$, finishing the result. □

To any ε-Rokhlin tower $R = \{H_i \times A_i\}_{i=1}^N$ for T, we can associate a **tower partition** Π_R by

$$\Pi_R = \begin{cases} (i, S(x)) & \text{if } x \in A_i, \\ (0, \emptyset) & \text{for all other } x. \end{cases}$$

Lemma 4.0.4. *For any* $\varepsilon > 0$ *and* N, *there exists* $\delta < \frac{1}{2}$ *such that if* $R = \{H_i \times A_i\}_{i=1}^N$ *is a* δ*-Rokhlin tower with* $\#H_i > \frac{1}{\delta}$, *for all* i, *then* $H(\Pi_R) < \varepsilon$.

Proof First notice that, for all i,

$$\mu(A_i) \le \frac{1}{\#H_i(1-\delta)} \le \frac{\delta}{1-\delta} < 2\delta.$$

Each A_i is partitioned by a choice for $S(x)$ into at most $\binom{\#H_i}{\delta\#H_i}$ sets. Thus

$$H(\Pi_R) \le -2\delta N \ln(2\delta) + \frac{1}{1-\delta}\sum_{i=1}^{N}\frac{1}{\#H_i}\ln\binom{\#H_i}{\delta\#H_i} \le \varepsilon,$$

if δ is chosen small enough. □

We now prove the first basic copying lemma. The proof follows fairly standard lines, but a number of parts in the statement are a bit different from the usual. We have tried to tie together in this one result all of the copying type arguments we will need. In the corollaries following the proof we will state the special cases we will use.

Theorem 4.0.5. *Suppose $(X,\mathscr{F},\mu,T)$ is a measure-preserving ergodic G-action and $P : X \to \Sigma_P$ and $Q : X \to \Sigma_Q$ are two finite partitions. For any $\varepsilon > 0$ there is a $\delta > 0$ satsifying the following. Suppose $(X_1,\mathscr{F}_1,\mu_1,T_1)$ is another measure preserving and ergodic G-action, and $Q_1 : X \to \Sigma_Q$ is a partition satisfying*

(1) $\|(T_1,Q_1),(T,Q)\|_* < \delta$.
Then there exists a partition $P' : X_1 \to \Sigma_P$ satisfying:
(1′) $\|(T,P\vee Q),(T_1,P'\vee Q_1)\|_* < \varepsilon$; *and*
(2′) $h(T_1,P') \ge (h(T_1,Q_1) - h(T,Q)) + \min\{h(T,P),h(T,Q)\} - \varepsilon$.

Proof Fix ε. Let $i_0 = [\log_2(2/\varepsilon)] + 1$ so that if

$$|(T,P\vee Q),(T_1,P'\vee Q_1)|_{F_{i_0}} < \varepsilon/2$$

we will have obtained (1′).

Let

$$a = \min\{\mu(f) : \mu(f) > 0 \text{ and } f \text{ is an atom of } \bigvee_{g\in F_{i_0}} T_{g^{-1}}(P\vee Q)\}.$$

Let n_0 be the number of elements in $\bigvee_{g\in F_{i_0}} T_{g^{-1}}(P\vee Q)$. Let

$$\bar{\varepsilon} = \left(\frac{\varepsilon}{10}\right)\frac{a}{n_0\#F_{i_0}}.$$

Choose $K \subseteq G$ with $F_{i_0} \subseteq K$, choose δ and N so that, applying Theorem 3.0.5, for any (δ,K)-invariant sequence $H_1,\ldots,H_N$, we can construct an $\bar{\varepsilon}$-Rokhlin tower, $R = \{H_i \times A_i\}_{i=1}^N$ in any free ergodic G-action (in particular, $(X_1, \mathscr{F}_1, \mu_1, T_1)$ once it is chosen).

Fixing N, we will place further requirements on the sets $H_1, \ldots, H_N$, as follows:

(1) for any $\bar{\varepsilon}$-Rokhlin tower $R = \{H_i \times A_i\}_{i=1}^N$, by Lemma 4.0.4,

$$H(\Pi_R) < \frac{\varepsilon}{10};$$

(2) for any atom $f \in \bigvee_{g \in F_{i_0}} T_{g^{-1}}(P \vee Q)$, by the mean ergodic theorem,

$$\| \frac{1}{\#H_i} \sum_{g \in H_i} 1_f T_g - \mu(f) \|_1 < \bar{\varepsilon}^3;$$

(3) letting

$$C_i(P) = \{f' \in \vee_{g \in H_i} T_{g^{-1}}(P) : \mu(f') = 2^{-(h(T,P) \pm \frac{\bar{\varepsilon}}{10})\#H_i}\} \text{ and}$$
$$C_i(Q) = \{f'' \in \vee_{g \in H_i} T_{g^{-1}}(Q) : \mu(f'') = 2^{-(h(T,Q) \pm \frac{\bar{\varepsilon}}{10})\#H_i}\},$$

the H_i are sufficiently invariant that the Shannon–McMillan theorem (Corollary 3.0.10) gives $\mu(C_i(P) \cap C_i(Q)) \geq 1 - \bar{\varepsilon}^2$; and

(4) the H_i are so invariant that

$$\frac{\#(F_{i_0} H_i \triangle H_i)}{\#H_i} < \bar{\varepsilon}^2.$$

Set

$$D_i = \left\{ \bar{f} \in \bigvee_{g \in H_i} T_{g^{-1}}(P \vee Q) : \text{ for some } x \in \bar{f} \text{ and all} \right.$$
$$\left. f \in \vee_{g \in F_{i_0}} T_{g^{-1}}(P \vee Q), \left| \frac{1}{\#H_i} \sum_{g \in H_i} 1_f(T_g(x)) - \mu(f) \right| < \bar{\varepsilon} \right\}.$$

By (2), above, we know that $\mu(\cup_{\bar{f} \in D_i} \bar{f}) > 1 - \bar{\varepsilon}^2$. Define

$$E_i = \left\{ f'' \in C_i(Q) : \mu(f'' \cap (\cup_{\bar{f} \in D_i} \bar{f})) \geq (1 - \bar{\varepsilon})\mu(f'') \text{ and} \right.$$
$$\left. \mu(f'' \cap (\cup_{f' \in C_i(P)} f')) \geq (1 - \bar{\varepsilon})\mu(f'') \right\}.$$

One calculates that $\mu(\cup_{f'' \in E_i} f'') > 1 - 3\bar{\varepsilon}$.

For $f'' \in E_i$, let

$$\eta(f'') = \{f' \in C_i(P) : \bar{f} = f'' \cap f' \in D_i\}.$$

We know that for each $f'' \in E_i$,

$$\mu(\bigcup_{f' \in \eta(f'')} (f' \cap f'')) > (1 - 2\bar{\varepsilon})\mu(f'').$$

We want to assign to each $f'' \in E_i$ an atom $f' \in \eta(f'')$. This assignment need not be 1-1, but we want to control how close it is to being 1-1.

Choose a model non-atomic measure space (M_i, ν) of total mass

$$\mu(\bigcup_{f'' \in E_i} (\bigcup_{f' \in \eta(f'')} (f' \cap f''))) > 1 - 3\bar{\varepsilon}.$$

Partition M_i into pieces labeled by sets $f' \cap f''$, $f'' \in E_i$, $f' \in \eta(f'')$, each of mass $\mu(f' \cap f'')$. In this model, a set f' is just a union of all sets labeled $f' \cap f''$, and similarly for f''.

Choose an integer N_i with

$$\log_2(N_i) = \left[h(T,Q) - \min(h(T,P), h(T,Q)) + \frac{2\bar{\varepsilon}}{5} \pm \frac{\bar{\varepsilon}}{10}\right] \#H_i.$$

(Note that this requires $\#H_i > 10/\varepsilon$.)

Refine each atom $f' \cap f''$ in the model into N_i sets of equal mass, labeled $\{1, \ldots, N_i\}$. Let f'_j be the union of pieces whose first label is f' and third label is $j \in \{1, \ldots, N_i\}$,

$$f'_j = \bigcup_{f'' \in E_i} (f' \cap f'' \cap j).$$

We calculate that

$$\begin{aligned} \nu(f'_j) &\leq 2^{-(h(T,P) \pm \frac{\bar{\varepsilon}}{10})\#H_i} N_i^{-1} \\ &\leq 2^{-(h(T,Q) + \frac{2\bar{\varepsilon}}{10})\#H_i} \\ &\leq \nu(f'')2^{-\frac{\bar{\varepsilon}}{10}\#H_i} \end{aligned}$$

for all $f'' \in C_i(Q)$.

Making sure that $\frac{\bar{\varepsilon}}{10}\#H_i > \log\left(\frac{1}{\bar{\varepsilon}}\right)$ we see that $\nu(f'_j) \leq \bar{\varepsilon}\,\nu(f'')$.

Select a 1-1 map

$$p : E_i \to \{f'_j : j = 1, 2, \ldots, N_i, \quad f' \in C_i(P)\}$$

with the condition that $p(f'') \cap f'' \neq \emptyset$ and such that $\nu(\text{Domain}(p))$ is of maximal size. Let $U_i = \text{Domain}(p)$.

We claim that $\nu(U_i) > 1 - 4\bar{\varepsilon}$. If not, then there is a collection

U_i^C of atoms $f'' \in E_i$, not in the domain of p, with $\nu(U_i^C) > \bar{\varepsilon}$. Let $V = \{f'_j \in \text{Range}(p)\}$. Since p is 1-1 and each $\nu(p(f'')) \le \bar{\varepsilon}\, m(f'')$, we see that $\nu(V) \le \bar{\varepsilon}$. Hence there must exist some $f'' \notin U_i$ and $f'_j \notin V$ such that $f'' \cap f'_j \neq \emptyset$ and p could be extended to f'' by setting $p(f'') = f'_j$. This contradicts the maximality of U_i completing the claim.

If $p(f'') = f'_j$ where $f'_j = \bigcup_{f'' \in E_i} (f' \cap f'' \cap j)$, then define

$$q(f'') = n_i(f') \in \Sigma_P^{H_i},$$

the T, Σ_P, H_i-name of f'.

We are now ready to select δ. For any free and ergodic G-action $(X_1, \mathscr{F}_1, \mu_1, T_1)$ and partition $Q_1 : X_1 \to \Sigma_Q$, for any $f'' \in \bigvee_{g \in H_i} T_{g^{-1}}(Q)$, there will be an atom $f''_1 \in \bigvee_{g \in H_i} T_{1,g^{-1}}(Q_1)$ with the same Σ_Q, H_i-name. Choose δ so small that $\|(T,Q),(T_1,Q_1)\|_* < \delta$ will imply that for all H_i, and all $f'' \in \bigvee_{g \in H_i} T_{g^{-1}}(Q)$ with $\mu(f'') > 0$ we will have

$$|\mu(f'') - \mu_1(f''_1)| < \bar{\varepsilon}\mu(f'').$$

Suppose $(X_1, \mathscr{F}_1, \mu_1, T_1)$ is a free and ergodic G-action and Q_1 a partition of X_1 satisfying (1) of Theorem 4.0.5.

Construct an $\bar{\varepsilon}$-Rokhlin tower $\{H_i \times A_i\}_{i=1}^N$ in X_1 such that

$$A_i \perp \bigvee_{g \in H_i} T_{1,g^{-1}}(Q_1).$$

Partition each A_i according to the atoms $f''_1 \in \bigvee_{g \in H_i} T_{1,g^{-1}}(Q_1)$, and assign to each $A_i \cap f''_1$ the name $q(f'')$. A fraction of at most $\bar{\varepsilon}$ of each A_i may lie in atoms f''_1 whose H_i, P-names do not occur in the (T,Q) process. Assign to them any P-name you wish, still calling the assignment $q(f'')$. Paint these names on the towers as described earlier. This gives a partition $P' : X_1 \to \Sigma_P$ of the tower image. Extend this partition outside the tower image, in any measurable fashion.

Further refine P' by also assigning to the points in a slice $S(x_1)$, $x_1 \in f''_1$, the index j if $p(f'') = f'_j$. Call this refinement

$$\hat{P} : X_1 \to \Sigma_P \times \{1, \ldots, \sup_{1 \le i \le N} (N_i)\}.$$

To verify conclusion (1′) consider the set

$$Z = \{x'_1 : \text{ for all } g \in F_{i_0}, T_{1,g}(x'_1) \in S(x_1), \text{ for } x_1 \in A_i \text{ with } x_1 \in f''_1 \text{ where } f'' \in (E_i \cap U_i)\}.$$

That is,

$$Z = \bigcup_{f'' \in U_i \cap E_i} \left(\bigcup_{x_1 \in f_1''} \big(\bigcap_{g \in F_{i_0}} T_{1,g^{-1}}(S(x))\big) \right).$$

As Z has measure within $\bar{\varepsilon}$ of the corresponding set in X,

$$\mu_1(Z) > 1 - 5\bar{\varepsilon}.$$

For any atom $\tilde{f} \in \bigvee_{g \in F_{i_0}} T_{1,g^{-1}}(P' \vee Q_1)$, let $f \in \bigvee_{g \in F_{i_0}} T_{g^{-1}}(P \vee Q)$ be the atom of the same name.

For $x_1' \in Z$ we then have that $x_1' \in \tilde{f}$ iff $x_1' = T_{1,g}(x_1)$, for some $g \in H_i, x_1 \in f_1''$, such that the name of $(f'', q(f'')) \in (\Sigma_Q \times \Sigma_P)^{H_i}$, in $(\Sigma_Q \times \Sigma_P)^{H_i}$ restricted to the indices $(F_{i_0} g)$ is the $T, \Sigma_Q \times \Sigma_P, F_{i_0}$-name of f.

For $x_1 \in f_1''$, $f'' \in U_i$ and $f \in \bigvee_{g \in F_{i_0}} T_{g^{-1}}(P \vee Q_1)$ we compute that

$$\begin{aligned}
\#\{x_1' \in \tilde{f} \cap S(x_1) : T_{1,g}(x') \in S(x_1) \text{ for all } g \in F_{i_0}\} &= \#H_i(\mu(f) \pm \bar{\varepsilon}) \\
&\pm \#\{g_0 \in H_i : T_1|_{H_i \times A_i} \text{ is not 1-1 at some } (gg_0, x_1), g \in F_{i_0}\} \\
&\pm \#\{g_0 \in H_i : gg_0 \notin H_i \text{ for some } g \in F_{i_0}\} \\
&= \mu(f)(1 \pm 3\bar{\varepsilon})\#H_i.
\end{aligned}$$

Since $\mu_1(Z) \geq 1 - 5\bar{\varepsilon} \geq 1 - \frac{3}{10}\varepsilon\mu(f)$, for all $f \in \bigvee_{g \in F_{i_0}} T_{g^{-1}}(P \vee Q_1)$, $|\mu(f) - \mu_1(\tilde{f})| \leq .6\varepsilon\mu(f)$, which is more than enough for (1′).

To obtain the entropy estimate (2′), we calculate a lower estimate for $h(T_1, \hat{P})$ and an upper estimate for

$$h(T_1, \hat{P}|P') = h(T_1, \hat{P} \vee P') - h(T_1, P').$$

Let $\tilde{P}$ be the partition of X_1 defined by

$$\tilde{P}(x_1) = \begin{cases} p(f) & \text{if } x_1 \in A_i \cap f_1, f \in U_i \\ \star & \text{otherwise.} \end{cases}$$

Now for $x_1 \in A_i \cap f_1$, $f \in U_i$, the $T_1, \hat{P}, H_i$-name of x_1 can differ from $p(f)$ only on $H_i \setminus g(x_1)$. Thus an atom of $\bigvee_{g \in H_i} T_{1,g^{-1}}(\hat{P})|_{A_i}$ contains at most $\#P^{\bar{\varepsilon}\#H_i}$ atoms of $\tilde{P}$. Hence we see that

$$H(\tilde{P}| \bigvee_{g \in G} T_{1,g^{-1}}(\Pi_R \vee \hat{P})) \leq \sum_{i=1}^{N} \mu_1(A_i)\bar{\varepsilon}\#H_i \log_2(\#P) \leq \frac{\varepsilon}{10},$$

since $\sum_{i=1}^{N} \#H_i\mu_1(A_i) < 1 + \bar{\varepsilon}$ and $\log_2(\#P) \leq \log(n_0) < n_0$.

Now Q_1 is measurable with respect to the span of

$$\bigvee_{g\in G} T_{1,g^{-1}}(\tilde{P}),\ \bigvee_{g\in G} T_{1,g^{-1}}(\Pi_R) \text{ and } \bigvee_{g\in G} T_{1,g^{-1}}(\overline{Q}),$$

where

$$\overline{Q}(x_1) = \begin{cases} Q_1(x_1) & \text{if } x_1 \text{ is outside the tower image} \\ \star & \text{if } x_1 \text{ is in the tower image.} \end{cases}$$

Since $h(T_1, \Pi_R) < \frac{\varepsilon}{10}$ and

$$h(T_1, \bar{Q}) \le H(\mu_1(\text{tower image})) + (1 - \mu_1(\text{tower image}))H(Q) \le \frac{\varepsilon}{10},$$

we see that $h(T_1, \tilde{P}) \ge h(T_1, Q_1) - \frac{3\varepsilon}{10}$, and so

$$h(T_1, \hat{P}) \ge h(T_1, Q) - \frac{2\varepsilon}{5}.$$

To obtain an upper estimate of $h(T_1, \hat{P}|P')$, let I be the partition defined by

$$I(x) = \begin{cases} i & \text{if } x_1 \in f_1'' \cap A_j, f'' \in U_i,\ p(f'') = f_i' \\ \star & \text{otherwise.} \end{cases}$$

Obviously $\hat{P}$ is $\bigvee_{g\in G} T_{1,g^{-1}}(P' \vee I)$-measurable, and so

$$h(T_1, \hat{P}|P') = h(T_1, \hat{P}) - h(T, P') \le h(T_1, I) \le H(I).$$

But

$$\begin{aligned} H(I) &\le H(\Pi_R) + \sum_{i=1}^{N} \mu_1(A_i)\log_2(N_i) \\ &\le \frac{\varepsilon}{10} + \sum_{i=1}^{N} (\mu(A_i)\#H_i)[h(T,Q) - \min(h(T,P), h(T,Q)) + \bar{\varepsilon}/2] \\ &\le \frac{2\varepsilon}{10}[h(T,Q) - \min(h(T,Q), h(T,P)]. \end{aligned}$$

Thus

$$\begin{aligned} h(T_1, P') &= h(T_1, \hat{P} \vee P') - h(T_1, \hat{P}|P') \\ &\ge h(T_1, \hat{P}) - h(T_1, \hat{P}|P') \\ &\ge h(T_1, Q_1) - \frac{2\varepsilon}{5} - \frac{\varepsilon}{5} - h(T,Q) + \min(h(T,Q), h(T,P)) \\ &> h(T_1, Q_1) - h(T,Q) + \min(h(T_1,Q_1), h(T,P)) - \varepsilon. \qquad \square \end{aligned}$$

We now state a version of Theorem 4.0.5 with an entropy bound on the process (T_1, Q_1).

Corollary 4.0.6. *Suppose $(X, \mathscr{F}, \mu, T)$ is a measure-preserving ergodic G-action and $P : X \to \Sigma_P$ and $Q : X \to \Sigma_Q$ are two finite partitions. For any $\varepsilon > 0$ there is a $\delta > 0$ satsifying the following. Suppose $(X_1, \mathscr{F}_1, \mu_1, T_1)$ is another measure preserving and ergodic G-action, and $Q_1 : X \to \Sigma_Q$ is a partition satisfying*

(1) $\|(T_1, Q_1), (T, Q)\|_* < \delta$ *and*
(2) $h(T_1, Q_1) > h(T, Q) - \frac{\varepsilon}{2}$.

Then there exists a partition $P' : X_1 \to \Sigma_P$ satisfying:

(1′) $\|(T, P \vee Q), (T_1, P' \vee Q_1)\|_* < \varepsilon$; *and*
(2′) $h(T_1, P') \geq \min\{h(T, P), h(T, Q)\} - \varepsilon$.

Proof Use $\varepsilon/2$ in Theorem 4.0.5, ensure that $\delta < \varepsilon/2$ and (2′) of Theorem 4.0.5 easily gives (2′) of the corollary. □

We now state two particular cases of this corollary of interest to us.

Corollary 4.0.7. *Suppose $(X, \mathscr{F}, \mu, T)$ is a free and ergodic G-action, $P : X \to \Sigma_P$ and $Q : X \to \Sigma_Q$ are two finite partitions. Suppose $\mathscr{H} \subseteq \mathscr{F}$ is a T-invariant sub-σ-algebra on which T still acts freely, and Q is $\mathscr{H}$-measurable.*

For any $\varepsilon > 0$ there exists a partition $P' : X \to \Sigma_P$ that is $\mathscr{H}$-measurable satisfying:

(1) $\|(T, P \vee Q), (T, P' \vee Q)\|_* < \varepsilon$; *and*
(2) $h(T, P') \geq \min(h(T, P), h(T, Q)) - \varepsilon$.

Proof Just let (T_1, Q_1) be (T, Q) restricted to the factor action on $\mathscr{H}$ in Corollary 4.0.6. □

Corollary 4.0.8. *Suppose $(X, \mathscr{F}, \mu, T)$ is a free and ergodic G-action and $P : X \to \Sigma_P$ and $Q : X \to \Sigma_Q$ are two finite partitions and $\varepsilon > 0$. Let $(Y, \mathscr{G}, \nu, S)$ be a Bernoulli action with $h(S) + h(T, Q) \geq h(T, P) - \varepsilon/3$. There exists a δ_0 (which we assume to be $< \varepsilon/3$) satisfying the following. Suppose $(X_1, \mathscr{F}_1, \mu_1, T_1)$ is any other free and ergodic G-action, and $Q_1 : X_1 \to \Sigma_Q$ is a finite partition with*

(1) $\|(T, Q), (T_1, Q_1)\|_* < \delta_0$ *and*
(2) $h(T_1, Q_1) \geq h(T, Q) - \varepsilon/6$.

Then there exists a partition $P' : X_1 \times Y \to \Sigma_P$ so that in the product action $(X_1 \times Y, \mathscr{F} \times \mathscr{G}, \mu_1 \times v, T_1 \times S)$ we conclude

(3) $\|(T_1 \times S, Q_1 \vee P'), (T, Q \vee P)\|_* < \varepsilon$ *and*
(4) $h(T_1 \times S, Q_1 \vee P') \geq h(T, Q \vee P) - \varepsilon$.

Proof Let $B : Y \to \Sigma_B$ be a finite and independent generator for the Bernoulli process. Consider the free and ergodic G-action $T \times S$ with respect to product measure $\mu \times v$. Consider on it the pair of partitions $\overline{P} : X \times Y \to \Sigma_P$, $\overline{P}(x,y) = P(x)$, and $\overline{Q} : X \times Y \to \Sigma_Q \times \Sigma_B$, $\overline{Q}(x,y) = (Q(x), B(y))$.

Notice that for any free and ergodic G-action T_1 with partition $Q_1 : X_1 \to \Sigma_Q$ we have

$$\|(T_1 \times S, Q_1 \times B), (T \times S, \overline{Q})\|_* = \|(T_1, Q_1), (T, Q)\|_*.$$

Thus applying Corollary 4.0.6 to the G-action $T \times S$ with partitions $\overline{P}$ and $\overline{Q}$, and $\varepsilon/3$, there exists $0 < \delta_0 < \varepsilon/3$ so that if

$$\|(T_1, Q_1), (T, Q)\|_* \leq \delta_0,$$

then there will exist a partition $P' : X_1 \times Y \to \Sigma_P$ with

$$\|(T_1 \times S, Q_1 \vee B \vee P'), (T \times S, \overline{Q} \vee \overline{P})\|_* < \varepsilon.$$

This distribution distance though is at least as large as

$$\|(T_1 \times S, Q_1 \vee P'), (T, Q \vee P)\|_*$$

which is (3). We also have from Corollary 4.0.6 that

$$h(T_1 \times S, P') \geq \min\{h(T, P), h(T_1 \times S, Q_1 \vee B)\} - \varepsilon/3.$$

As

$$h(T_1 \times S, Q_1 \vee B) \geq h(T \times S, Q \vee B) - \delta \geq h(T, P) - 2\varepsilon/3,$$

this latter minimum is at least $h(T, P) - \varepsilon$ giving us (4) and the result. □

The following Corollary to Theorem 4.0.5 states that if the partition P had arisen from a rearrangement then the new partition P' could also be chosen to have arisen from a rearrangement.

Corollary 4.0.9. *Let (α, ϕ) be any G-rearrangement. Suppose $Q : X \to \Sigma_Q$*

is a finite partition. For any $\varepsilon > 0$, there exists $\delta > 0$ so that the following holds. Suppose $(X_1, \mathscr{F}_1, \mu_1, T_1^{\alpha_1})$ is a free and ergodic G-action, and $Q_1 : X_1 \to \Sigma_Q$ is a finite partition satisfying

$$\|(T^\alpha, Q), (T_1^{\alpha_1}, Q_1)\|_* < \delta.$$

Then there exists ϕ' in the full-group of $T_1^{\alpha_1}$ such that

$$\|(T^\alpha, g_{(\alpha,\phi)} \vee Q), (T_1^{\alpha_1}, g_{(\alpha_1,\phi')} \vee Q_1)\|_* < \varepsilon.$$

Proof By Theorem 4.0.2 we can assume that (α, ϕ) is a bounded rearrangement. Hence $g_{(\alpha,\phi)} : X \to G$ given by $g_{(\alpha,\phi)}(x) = \alpha(x, \phi(x))$ is a finite partition. Let $\Sigma_\phi \subseteq G$ denote its set of labels.

Let $\varepsilon > 0$. Notice that if $\varepsilon' < \dfrac{\varepsilon}{2\#F_i}$ where $i > \log_2(2/\varepsilon)$ then for any partitions R and R' with $\|R, R'\|_1 < \varepsilon'$ we will have $\|R, R'\|_* < \varepsilon/2$.

Apply Theorem 4.0.3 to obtain a $\delta_1 < \varepsilon'$ so that if

$$\|(T^\alpha, g_{(\alpha,\phi)}), (T_1^{\alpha_1}, P')\|_* < \delta_1$$

then there will be a ϕ' in the full-group of $T_1^{\alpha_1}$ with

$$\|P', g_{(\alpha_1,\phi')}\|_1 < \varepsilon/2.$$

Apply Theorem 4.0.5 to see that there exists a $\delta < \varepsilon/2$ so that for any free and ergodic $(X_1, \mathscr{F}_1, \mu_1, T_1^{\alpha_1})$ and finite partition $Q_1 : X_1 \to \Sigma_Q$ satisfying

$$\|(T^\alpha, Q), (T_1^{\alpha_1}, Q_1)\|_* < \delta$$

then there will be a partition $P' : X_1 \to \Sigma_\phi \subseteq G$ such that

$$\|(T^\alpha, g_{(\alpha,\phi)} \vee Q), (T_1^{\alpha_1}, P' \vee Q_1)\|_* < \delta_1.$$

Theorem 4.0.3 now tells us there is a ϕ' in the full-group of $T_1^{\alpha_1}$ with

$$\|(T_1^{\alpha_1}, P' \vee Q_1), (T_1^{\alpha_1}, g_{(\alpha_1,\phi')} \vee Q_1)\|_1 < \varepsilon'$$

and hence

$$\|(T_1^{\alpha_1}, P' \vee Q_1), (T_1^{\alpha_1}, g_{(\alpha_1,\phi')} \vee Q_1)\|_* < \varepsilon/2$$

and the result follows. □

Corollary 4.0.10. *Suppose $(X, \mathscr{F}, \mu, T^\alpha)$ is a free and ergodic G-action, $Q : X \to \Sigma_Q$ is a finite partition and ϕ is in the full-group of T^α. Given any $\varepsilon > 0$ there is a $\delta > 0$ satisfying the following:*

Suppose $(X_1, \mathscr{F}_1, \mu_1, T_1^{\alpha_1})$ is another free and ergodic G-action, and $Q_1 : X_1 \to \Sigma_Q$ is a finite partition satisfying

(1) $\|(T^{\alpha}, Q), (T_1^{\alpha_1}, Q_1)\|_* < \delta$ *and*
(2) $h(T^{\alpha_1}, Q_1) \geq h(T^{\alpha}, Q) - \varepsilon/6$.

For $(Y, \mathcal{G}, \nu, S)$ *a Bernoulli G-action with*

$$h(S) + h(T, Q) \geq h(T^{\alpha\phi}, Q) - \varepsilon/3,$$

let $U^{\alpha_1} = (T^{\alpha_1} \times S)$ *be the direct product action. Then there exists* ϕ' *in the full-group of* U^{α_1} *such that*

(3) $\|(U^{\alpha_1\phi'}, Q_1), (T^{\alpha\phi}, Q)\|_* < \varepsilon$,
(4) $\|(\alpha_1, \phi'), (\alpha, \phi)\|_* < \varepsilon$, *and*
(5) $h(U^{\alpha_1\phi}, Q_1) > h(T^{\alpha\phi}, Q) - \varepsilon$.

Proof To begin, notice that $(T^{\alpha\phi}, Q)$ and $(T^{\alpha}, Q \circ \phi^{-1})$ are identical in distribution.

Furthermore, notice that, given any $\varepsilon_0 > 0$ (whose value we will set later), there exists $\varepsilon_1 > 0$ so that if

$$\|(U^{\alpha_1}, Q_1 \vee Q' \vee g_{(\alpha_1,\phi')}), (T^{\alpha}, Q \vee (Q \circ \phi^{-1}) \vee g_{(\alpha,\phi)})\|_* < \varepsilon_1$$

then

$$(\mu_1 \times \nu)(Q_1 \circ \phi'^{-1} \triangle Q') < \varepsilon_0.$$

This is simply the observation (explored in more detail in Lemma 4.0.11 below) that there is a coding from the $Q \vee g_{(\alpha,\phi)}$-name of a point to its $Q \circ \phi^{-1}$-name, and applying this coding to the $Q_1 \vee g_{(\alpha_1,\phi')}$-name will construct the $Q_1 \circ \phi'^{-1}$-name, which must agree most of the time with the Q'-name, if ε_1 is small.

Given $\varepsilon_2 > 0$ whose value will be set later, choose δ with $\varepsilon_2 > \delta > 0$ so small that (1) and (2) (using $P = Q \circ \phi^{-1}$ in Corollary 4.0.8) imply there is a partition $Q' : X_1 \times Y \to \Sigma_Q$ with

(1′) $\|(U^{\alpha_1}, Q_1 \vee Q'), (T^{\alpha}, Q \vee (Q \circ \phi^{-1}))\|_* < \varepsilon_2$ and
(2′) $h(U^{\alpha_1}, Q') \geq h(T^{\alpha}, Q \circ \phi^{-1}) - 2\varepsilon_2 = h(T^{\alpha\phi}, Q) - 2\varepsilon_2$.

Given ε_1 from above, there is an ε_2 so that (1′) and Corollary 4.0.9 imply there is a ϕ' in the full-group of U^{α_1} with

$$\|(U^{\alpha_1}, Q_1 \vee Q' \vee g_{(\alpha_1,\phi')}), (T^{\alpha}, Q \vee (Q \circ \phi^{-1}) \vee g_{(\alpha,\phi)})\|_* < \varepsilon_1$$

and hence

$$\|Q_1 \circ \phi'^{-1}, Q'\|_1 < \varepsilon_0.$$

If ε_0 is small enough, then

$$\begin{aligned}
&\|(U^{\alpha_1}, Q_1 \circ \phi'^{-1}), (T^{\alpha}, Q \circ \phi^{-1})\|_* \\
&\leq \|(U^{\alpha_1}, Q_1 \circ \phi'^{-1}), (U^{\alpha_1}, Q')\|_* + \|(T_1^{\alpha_1}, Q'), (T^{\alpha}, Q \circ \phi^{-1})\|_* \\
&< \|(U^{\alpha_1}, Q_1 \circ \phi'^{-1}), (U^{\alpha_1}, Q')\|_* + \varepsilon_2
\end{aligned}$$

can be made $< \varepsilon$ (which is (3)) and

$$\begin{aligned}
h(U^{\alpha_1\phi'}, Q_1) &= h(U^{\alpha_1}, Q_1 \circ \phi'^{-1}) \\
&\geq h(U^{\alpha_1}, Q') - \varepsilon_0 \log(\#\Sigma_Q) - H(\varepsilon_0) \\
&\geq h(T^{\alpha\phi}, Q) - 2\varepsilon_2 - \varepsilon_0 \log(\#\Sigma_Q) - H(\varepsilon_0)
\end{aligned}$$

can be made $\geq h(T^{\alpha\phi}, Q) - \varepsilon$ (which is (5)).

Lastly, as

$$\|(U^{\alpha_1}, g_{(\alpha_1,\phi')}), (T^{\alpha}, g_{(\alpha,\phi)})\|_* < \varepsilon_1,$$

if ε_1 is small enough,

$$\|(\alpha_1, \phi'), (\alpha, \phi)\|_* < \varepsilon$$

which is (4). □

Lemma 4.0.11. *Suppose $(X, \mathcal{F}, \mu, T^{\alpha})$ is a free ergodic G-action, ϕ is in its full-group with $g_{(\alpha,\phi)}$ finite-valued (that is to say (α, ϕ) is a bounded rearrangement) and $Q : X \to \Sigma_Q$ is a finite partition. Given any $\varepsilon > 0$ there is a $\delta > 0$ so that in any other free and ergodic G-action $(X_1, \mathcal{F}_1, \mu_1, T_1^{\alpha_1})$ with*

(1) $\|(T^{\alpha}, Q \vee g_{(\alpha,\phi)}), (T_1^{\alpha_1}, Q_1 \vee g_{(\alpha_1,\phi_1)})\|_* < \delta$ *we will have*
(2) $\|(T^{\alpha\phi}, Q \vee g_{(\alpha\phi,\phi^{-1})}), (T_1^{\alpha_1\phi_1}, Q_1 \vee g_{(\alpha_1\phi_1,\phi_1^{-1})})\|_* < \varepsilon$.

Proof Remember that in general the two processes $(S^{\beta\psi}, R \circ \psi)$ and (S^{β}, R) are identical in distribution and so (2) can be written

$$\|(T^{\alpha}, (Q \vee g_{(\alpha\phi,\phi^{-1})}) \circ \phi^{-1}), (T_1^{\alpha_1}, (Q_1 \vee g_{(\alpha_1\phi_1,\phi_1^{-1})}) \circ \phi_1^{-1})\|_* < \varepsilon.$$

We now show that the $T^{\alpha}, (Q \vee g_{(\alpha\phi,\phi^{-1})}) \circ \phi^{-1}$-name of a point is a bounded code from its $T^{\alpha}, Q \vee g_{(\alpha,\phi)}$-name. That is to say there is a finite set $K \subseteq G$ with $(Q \vee g_{(\alpha\phi,\phi^{-1})}) \circ \phi^{-1}$ measurable with respect to $\bigvee_{g \in K} T^{\alpha}_{g^{-1}}(Q \vee g_{(\alpha,\phi)})$. We complete the result by noting that one now can apply this same code to the process $(T_1^{\alpha_1}, Q_1 \vee g_{(\alpha_1,\phi_1)})$ to obtain the partition $(Q_1 \vee g_{(\alpha_1\phi_1,\phi_1^{-1})}) \circ \phi_1^{-1}$ and the result.

We see this finite code by first noting

$$g_{(\alpha\phi,\phi^{-1})} \circ \phi^{-1}(x) = \alpha(x, \phi^{-1}(x)) = (g_{(\alpha,\phi)}(\phi^{-1}(x)))^{-1}$$

where $\phi^{-1}(x) = T^{\alpha}_{z(x)}(x)$ where $g_{(\alpha,\phi)}(\phi^{-1}(x)) = z(x)^{-1}$. This value $z(x)$ is a finite G-valued partition, taking values in the inverses of the values taken by $g_{(\alpha,\phi)}$. Thus z is obtained as a finite coding from the $T^{\alpha}, g_{(\alpha,\phi)}$-name of x. By searching out over this finite list of values for the solution to $g_{(\alpha,\phi)}(T^{\alpha}_{z}(x)) = z^{-1}$. Thus $g_{(\alpha\phi,\phi^{-1})} \circ \phi^{-1}(x)$ is a bounded code of the $T^{\alpha}, g_{(\alpha,\phi)}$-name of x.

As $Q \circ \phi^{-1}(x) = Q(T^{\alpha}_{z(x)}(x))$, and z is bounded, this is a finite code from the $T^{\alpha}, Q \vee g_{(\alpha,\phi)}$-name of x as well.

Notice that the code to $z(x)$ is canonical in that its dependence is really just on the function $f^{\alpha,\phi}_{x}|_K$ and not on the measure space giving rise to this particular function (see Lemma 6.3.5 for more on this). Hence precisely the same code will apply to the $T^{\alpha_1}_1, Q_1 \vee g_{(\alpha_1,\phi_1)}$-names. □

We now state a rather complicated result that is really little more than a series of observations from our current vantage, but states exactly what we will need in the equivalence theorem.

Theorem 4.0.12. *Suppose $(X, \mathcal{F}, \mu, T^{\alpha})$ is a free and ergodic G-action, P and Q are finite partitions. Suppose $\mathcal{H}$ is a T^{α}-invariant σ-algebra on which the action of T^{α} is still free, and Q is $\mathcal{H}$-measurable. Suppose also that $\phi_1, \phi_2, \ldots, \phi_I, \phi_{I+1}$ are all finite-valued elements in the full-group of T^{α} and that*

$$h(T^{\alpha}, Q) > h(T^{\alpha\phi_{I+1}}, P) - \varepsilon/2.$$

Then there exist full-group elements $\phi'_1, \ldots, \phi'_{I+1}$ and a partition P', all $\mathcal{H}$-measurable so that:

(1) $\|(T^{\alpha}, P \vee Q \vee \bigvee_{i=1}^{I+1} g_{(\alpha,\phi_i)}), (T^{\alpha}, P' \vee Q \vee \bigvee_{i=1}^{I+1} g_{(\alpha,\phi'_i)})\|_* < \varepsilon$; *and*

(2) $h(T^{\alpha\phi'_{I+1}}, P') \geq h(T^{\alpha\phi_{I+1}}, P) - \varepsilon$.

Proof We will work by first copying ϕ_{I+1}, then the partition P and then (the easy part) the rest of the full-group elements $\phi_1, \ldots, \phi_I$. For $\delta_1 > 0$ to be chosen later, if $\delta > 0$ is small enough, apply Corollary 4.0.9 to construct a full-group element ϕ'_{I+1} that is $\mathcal{H}$ measurable with

$$\|(T^{\alpha}, Q \vee g_{(\alpha,\phi_{I+1})}), (T^{\alpha}, Q \vee g_{(\alpha,\phi'_{I+1})})\|_* < \delta_1.$$

Using Lemma 4.0.11 for any $\delta_2 > 0$, if δ_1 is small enough we will have

$$\|(T^{\alpha\phi_{I+1}}, Q \vee Q \circ \phi_{I+1} \vee g_{(\alpha\phi_{I+1}, \phi_{I+1}^{-1})}),$$
$$(T^{\alpha\phi'_{I+1}}, Q \vee Q \circ \phi'_{I+1} \vee g_{(\alpha\phi'_{I+1}, \phi'^{-1}_{I+1})})\|_* < \delta_2.$$

Remembering that $(T^{\alpha\psi}, Q \circ \psi)$ and (T^α, Q) are identical in distribution, we know that

$$h(T^{\alpha\phi'_{I+1}}, Q \circ \phi'_{I+1}) > h(T^{\alpha\phi_{I+1}}, P) - \varepsilon/2.$$

Now applying Corollary 4.0.6 with

$$\begin{aligned}
\text{“}T_1\text{”} &= T^{\alpha\phi'_{I+1}}, \\
\text{“}Q_1\text{”} &= Q \vee Q \circ \phi'_{I+1} \vee g_{(\alpha\phi'_{I+1}, \phi'^{-1}_{I+1})}, \\
\text{“}T\text{”} &= T^{\alpha\phi_{I+1}}, \\
\text{“}Q\text{”} &= Q \vee Q \circ \phi_{I+1} \vee g_{(\alpha\phi_{I+1}, \phi_{I+1}^{-1})} \text{ and} \\
\text{“}P\text{”} &= P,
\end{aligned}$$

we can copy P into $\mathscr{H}$ as a partition P'. Thus for any $\varepsilon > \delta_3 > 0$ if δ_2 is small enough, we will obtain

$$\|(T^{\alpha\phi_{I+1}}, (P \vee Q \vee g_{(\alpha\phi_{I+1}, \phi_{I+1}^{-1})}) \circ \phi_{I+1}^{-1}),$$
$$(T_1^{\alpha_1\phi'_{I+1}}, (P' \vee Q_1 \vee g_{(\alpha_1\phi'_{I+1}, \phi'^{-1}_{I+1})}) \circ \phi'^{-1}_{I+1})\|_* < \delta_3.$$

More importantly though we get the entropy estimate:

$$h(T^{\alpha\phi'_{I+1}}, P') \geq \min\{h(T^{\alpha\phi_{I+1}}, P), h(T^{\alpha\phi'_{I+1}}, Q \circ \phi'_{I+1})\} - \varepsilon/2.$$

As

$$h(T^{\alpha\phi'_{I+1}}, Q \circ \phi'_{I+1}) = h(T^\alpha, Q) \geq h(T^{\alpha\phi_{I+1}}, P)$$

we obtain (2).

Continuing, for any $\delta_4 > 0$ if δ_3 was chosen small enough, by once more applying Lemma 4.0.11 we can pull back into the T^α and $T_1^{\alpha_1}$ systems with

$$\|(T^\alpha, P \vee Q \vee g_{(\alpha, \phi_{I+1})}), (T^\alpha, P' \vee Q \vee g_{(\alpha, \phi'_{I+1})})\|_* < \delta_4.$$

Now for the easy part. In Corollary 4.0.7, with “P” $= \bigvee_{i=1}^{I} g_{(\alpha,\phi_i)}$, for any $\delta_5 > 0$ if δ_4 is sufficiently small, we can construct a copy of the form

$\bigvee_{i=1}^{I} P_i'$ with

$$\|(T^{\alpha}, P \vee Q \vee g_{(\alpha,\phi_{I+1})} \vee \bigvee_{i=1}^{I} g_{(\alpha,\phi_i)}),$$

$$(T^{\alpha_1}, P' \vee Q_1 \vee g_{(\alpha_1,\phi'_{I+1})} \vee \bigvee_{i=1}^{I} P_i')\|_* < \delta_5.$$

If δ_5 is small enough, applying Theorem 4.0.3 to each of the P_i' we can modify them to become $g_{(\alpha_1,\phi_i')}$ obtaining (1) of the Theorem. □

We now develop a copying lemma that moves in precisely the opposite direction from Theorem 4.0.5 and its corollaries. There the principle intent was to maintain entropy as high as possible. Here it will be to decrease entropy as much as possible. This will play an essential role in our work on m-entropy. The proof follows exactly the same lines as that of Theorem 4.0.5 except that where we worked to make the assignment of names q as close to 1-1 as possible, here we will work to make it as far from 1-1 as possible.

Theorem 4.0.13. *Suppose $(X, \mathscr{F}, \mu, T)$ is a free and ergodic G-action and P and Q are two finite partitions and $\mathscr{H}$ is a T invariant sub-σ algebra of $\mathscr{F}$ on which the action is still free and for which Q is $\mathscr{H}$-measurable. Suppose*

(1) $h(T,Q|P) = h(T,Q \vee P) - h(T,P) > e > 0$. *Then for any* $\varepsilon > 0$ *there is a partition* $P' \in \mathscr{H}$ *with*
(2) $\|(T,P \vee Q),(T,P' \vee Q)\|_* < \varepsilon$ *and*
(3) $h(T,Q) - h(T,P') > e$.

Proof The proof is so parallel to that of Theorem 4.0.5 that when appropriate we will simply refer to the analogous facts in that proof.

Fix $\varepsilon > 0$ with $\varepsilon < h(T,P \vee Q) - h(T,Q)$ and let $i_0 = [\log_2(2/\varepsilon)] + 1$. We will obtain

$$|(T,P \vee Q),(T,P' \vee Q)|_{F_{i_0}} < \varepsilon/2$$

implying (2). (See the beginning of this section for definitions.)

Let

$$a = \min(\{\mu(f) : \mu(f) > 0 \text{ and } f \text{ is an atom of } \vee_{g \in F_{i_0}} T_{g^{-1}}(P \vee Q)\})$$

and let n_0 be the number of elements in $\bigvee_{g\in F_{i_0}} T_{g^{-1}}(P \vee Q)$. Let

$$\bar{\varepsilon} \leq \frac{\varepsilon a}{10 n_0 \# F_{i_0}}$$

be so small that

$$\binom{N}{\bar{\varepsilon} N} < 2^{N\varepsilon/10}.$$

Choose a finite subset $K \subseteq G$ with $F_{i_0} \subseteq K$ and values $\delta > 0$ and N so that Theorem 3.0.5 will guarantee that for any (δ, K)-invariant sequence of sets $H_1, \ldots, H_N$ we can construct an $\bar{\varepsilon}$-Rokhlin tower $R = \{H_i \times A_i\}_{i=1}^N$ in any free and ergodic G-action.

Fixing N we will place further requirements on the sets $H_1, \ldots, H_N$, as follows:

(i) for any $\bar{\varepsilon}$-Rokhlin tower $R = \{H_i \times A_i\}_{i=1}^N$, using Lemma 4.0.4 we want

$$H(\Pi_R) < \varepsilon/10;$$

(ii) for any atom $f \in \bigvee_{g\in F_{i_0}} T_{g^{-1}}(P \vee Q)$, by the mean ergodic theorem,

$$\left\| \frac{1}{\# H_i} \sum_{g\in H_i} 1_f T_g - \mu(f) \right\|_1 < \bar{\varepsilon}^5;$$

(iii) letting

$$C_i(P \vee Q) = \{\bar{f} \in \vee_{g\in H_i} T_{g^{-1}}(P \vee Q) : \mu(\bar{f}) = 2^{-(h(T,P\vee Q)\pm\varepsilon/10)\# H_i}\},$$

and

$$C_i(P) = \{f' \in \vee_{g\in H_i} T_{g^{-1}}(P) : \mu(f') = 2^{-(h(T,P)\pm\varepsilon/10)\# H_i}\},$$

the H_i are sufficiently invariant that the Shannon–McMillan theorem (Corollary 3.0.10) gives

$$\mu(C_i(P \vee Q) \cap C_i(P)) \geq 1 - \bar{\varepsilon}^4;$$

and

(iv) the H_i are so large and invariant that

$$\frac{\#(F_{i_0} h_i \triangle H_i)}{\# H_i} < \bar{\varepsilon}^2 \text{ and } 2^{-\# H_i \varepsilon/10} < \bar{\varepsilon}.$$

Set

$$D_i = \left\{ \overline{f} \in \bigvee_{g \in H_i} T_{g^{-1}}(P \vee Q) : \text{ for some } x \in \overline{f} \quad \text{and all} \right.$$

$$\left. f \in \bigvee_{g \in F_{i_0}} T_{g^{-1}}(P \vee Q), \left| \frac{1}{\# H_i} \sum_{g \in H_i} 1_f(T_g(x)) - \mu(f) \right| < \bar{\varepsilon} \right\}.$$

By (ii) we know

$$\mu(\bigcup_{\overline{f} \in D_i} \overline{f}) > 1 - \bar{\varepsilon}^4.$$

Define

$$E_i = \{ f'' \in \vee_{g \in H_i} T_{g^{-1}}(Q) :$$
$$\mu(f'' \cap (\cup_{\overline{f} \in D_i \cap C_i(P \vee Q)} \cap (\cup_{f' \in C_i(P)})) > (1 - 2\bar{\varepsilon}^2)\mu(f'')\}.$$

One calculates

$$\mu(\cup_{f'' \in E_i} f'') > 1 - 2\bar{\varepsilon}^2.$$

For $f'' \in E_i$ let

$$\eta(f'') = \{ f' \in C_i(P) : \overline{f} = f'' \cap f' \in D_i \cap C_i(P \vee Q)\}.$$

Hence for each $f' \in E_i$

$$\mu(\cup_{f' \in \eta(f'')}(f' \cap f'')) > (1 - 2\bar{\varepsilon}^2)\mu(f'')$$

and setting

$$G_i = \cup_{f'' \in E_i}(\cup_{f' \in \eta(f'')}(f'' \cap f''))$$

we have

$$\mu(G_i) > 1 - 4\bar{\varepsilon}^2.$$

Finally let

$$G_i(P) = \{ f' \in C_i(P) : \mu(f' \cap G_i) > (1 - 2\bar{\varepsilon})\mu(f')\}$$

and one calculates

$$\mu(G_i(P)) \geq (1 - 2\bar{\varepsilon}) - \bar{\varepsilon}^4 \geq 1 - 3\bar{\varepsilon}.$$

For any $f' \in G_i(P)$ notice that

$$\begin{aligned}\#\{f'' : f' \cap f'' \subseteq G_i\} &\geq \frac{(1-3\bar{\varepsilon})(2^{-(h(T,P)+\varepsilon/10)\#H_i})}{(2^{-(h(T,P\vee Q)-\varepsilon/10)\#H_i})}\\ &= (1-3\bar{\varepsilon})2^{(h(T,P\vee Q)-h(T,P)-\varepsilon/5)\#H_i}\\ &\geq (1-3\bar{\varepsilon})2^{(e+4\varepsilon/5)\#H_i}.\end{aligned}$$

As in Theorem 4.0.5 attempt to make an assignment

$$p : E_i \to G_i(P)$$

with $p(f'') \in \eta(f'')$ but here we push p as far away from one-to-one as possible. More precisely, suppose inductively p has been defined on a collection of atoms $D \subseteq E_i$. For the first step just assume D is empty.

If $\mu(D) > 1 - 4\bar{\varepsilon}$ stop. If not, then we must have

$$\mu((\cup_{f''\in D} f'')^c \cap (\cup_{f'\in G_i(P)} f')) \geq \bar{\varepsilon}$$

and for some $f' \in G_i(P)$ we must have

$$\mu(f' \cap (\cup_{f''\in D} f'')^c \cap (\cup_{f'\in G_i(P)} f')) \geq \bar{\varepsilon}\mu(f').$$

Hence for this f' we must have

$$\begin{aligned}\#\{f'' : f' \cap f'' \in G_i, \quad f'' \notin D\} &\geq \frac{\bar{\varepsilon}(2^{-(h(T,P)+\varepsilon/10)\#H_i})}{2^{-(h(T,P\vee Q)-\varepsilon/10)\#H_i}}\\ &\geq \bar{\varepsilon}\, 2^{(h(T,P\vee Q)-h(T,P)-\varepsilon/5)\#H_i}\\ &\geq \bar{\varepsilon}\, 2^{(e+4\varepsilon/5)\#H_i} > 2^{(e+7\varepsilon/10)\#H_i}.\end{aligned}$$

For this f' and any f'' with $f' \cap f'' \in G_i$ but $f \notin D$ we will have $f' \in \eta(f'')$. Extend p to these f'' by setting

$$p(f'') = f'.$$

Notice that this forces p to always be at least $\left(2^{(e+7\varepsilon/10)\#H_i}\right)$-to-one where it is defined. This assignment can proceed until $\mu(D) \geq 1 - 4\bar{\varepsilon}$.

We now proceed as in Theorem 4.0.5. Construct an $\bar{\varepsilon}$-Rokhlin tower $R = \{H_i \times A_i\}_{i=1}^N$ measurable with respect to the sub-algebra $\mathscr{H}$ with

$$A_i \perp \vee_{g\in H_i} T_{g^{-1}}(Q).$$

Partition each A_i according to the atoms $f'' \in \vee_{g\in H_i} T_{g^{-1}}(Q)$. For any $f'' \in D$, we have assigned a name $p(f'')$. For any $f'' \notin D$, assign to f'' a single name constant at all indices. Paint these P-names on the tower as described at the beginning of the section to construct the partition P'.

Outside of the tower image assign all points to the same symbol of Σ_P as was used to assign names to $f'' \notin D$.

That (2) now holds for $(T, P' \vee Q)$ is precisely the same as in the proof of Theorem 4.0.5 so we do not repeat it here. All we need argue is that

$$h(T, P') < h(T, Q) - e.$$

To see this we construct an auxiliary partition that encodes P'. Partition each A_i according to $\bigvee_{g \in H_i} T_{g^{-1}}(P')$. Call this partition W_i. Notice that as the map p is $\left(2^{-(e+7\varepsilon/10)\#H_i}\right)$ to 1 where it was originally defined, each slice $S(x)$ occupies all but at most $\overline{\varepsilon}\#H_i$ of H_i, and all points not in a slice over a name $f'' \in D$ are assigned the same symbol in Σ_P. Thus each partition W_i contains at most

$$\left(2^{-(e+7\varepsilon/10)} 2^{(h(T,Q)+\varepsilon/10)}\right)^{\#H_i} \binom{\#H_i}{\overline{\varepsilon}\#H_i}$$

elements. This value is of course

$$\leq 2^{(h(T,Q)-e-\varepsilon/5)\#H_i}.$$

Let W be the partition

$$W = \Pi_R \vee \begin{cases} W_i & \text{on} \quad Q_i \\ \star & \text{otherwise} \end{cases}.$$

It is easily seen that P' is measurable with respect to $\mathscr{H}(T, W)$. Hence

$$\begin{aligned} h(T, P') &\leq H(W) \\ &\leq \sum_i \mu(A_i)(h(T, Q) - e - \varepsilon/5)\#H_i + H(\Pi_R) \\ &\leq h(T, Q) - e - \varepsilon/10 < h(T, Q) - e. \end{aligned}$$

□

Corollary 4.0.14. *Suppose $(T, \mathscr{F}, \mu, T^\alpha)$ is a free and ergodic G-action, Q is a finite partition of X and ϕ is in the full-group of T^α satisfying*

(1) $h(T^\alpha, Q | Q \circ \phi^{-1}) = h(T, Q \vee (Q \circ \phi^{-1})) - h(T, Q \circ \phi^{-1}) > e.$

For any $\varepsilon > 0$ there exists ϕ' in the full-group of T^α with

(2) $\|(T^{\alpha\phi}, Q), (T^{\alpha\phi'}, Q)\|_* < \varepsilon$
(3) $\|(\alpha, \phi), (\alpha, \phi')\|_* < \varepsilon$ *and*
(4) $h(T^\alpha, Q \circ \phi'^{-1}) = h(T^{\alpha\phi}, P) < h(T, Q) - e.$

Proof Apply the argument in Lemma 4.0.11 to Theorem 4.0.13 instead of Theorem 4.0.5. As we are after an upper bound on the entropy of $h(T^\alpha, Q\circ\phi^{-1})$ as opposed to a lower bound, as obtained in Corollary 4.0.8, there is no need for the extra Bernoulli shift. Of course we set $T_1 = T$, $Q = Q_1$ and $P = Q \circ \phi^{-1}$. □

5
m-entropy

We define the m-entropy of T^α to be the infimum of entropies of T^β, over arrangements β in the m-equivalence class of α. Although our earlier work has made it essentially obvious, we will show that m-entropy is upper semi-continuous in m_α and hence this infimum is obtained on a dense G_δ subset of the equivalence class $E_m(\alpha)$. As in earlier work [25], [43], the main goal of this section is to show that for a fixed size m, either the m-entropy of T^α is zero, for all arrangements α, or the m-entropy is the usual entropy for all arrangements α. We will say that each size m is either entropy-free or entropy-preserving.

Let m be a fixed size. Let α be a G-arrangement. Recall that $E_m(\alpha) = \{\beta|\alpha \overset{m}{\sim} \beta\}$ is the m-equivalence class of α.

Definition 5.0.1. *Define the* **m-entropy** *of T^α to be*

$$h_m(T^\alpha) = \inf\{h(T^\beta); \beta \in E_m(\alpha)\}.$$

We first argue that this m-entropy is attained residually in $E_m(\alpha)$.

The following lemma tells us that if two arrangements are m-close then they are, in fact, close in distribution.

Lemma 5.0.2. *For every $\varepsilon > 0$ there exists $\delta > 0$ such that if*

$$m(\alpha, \beta) < \delta$$

then, for any partition P,

$$\|(T^\alpha, P), (T^\beta, P)\|_* < \varepsilon.$$

Proof This follows from Lemma 2.2.9 as if $P(T_g^\alpha(x)) \neq P(T_g^\beta(x))$ then in particular $\alpha(x, T_g^\beta(x)) \neq g$. □

Corollary 5.0.3. *Fix any G-arrangement α and partition P. Let $\varepsilon > 0$. There exists $\delta > 0$ such that if $m(\alpha, \beta) < \delta$ (i.e. if $\beta \in B_\delta(\alpha)$), then $h(T^\beta, P) < h(T^\alpha, P) + \varepsilon$.*

Proof This is a direct consequence of Lemma 5.0.2 and Theorem 3.0.17. □

Theorem 5.0.4. *The set of $\beta \in E_m(\alpha)$ with $h(T^\beta) = h_m(T^\alpha)$ is a dense G_δ in $E_m(\alpha)$. As $E_m(\alpha)$ is Polish, this infimum is achieved residually.*

Proof Fix α and P. Let

$$\mathcal{O}_{\varepsilon,P} = \{\beta \in E_m(\alpha); h(T^\beta, P) < h_m(T^\alpha) + \varepsilon\}.$$

If $h(T^\beta) = h_m(T^\alpha)$ then $\beta \in \mathcal{O}_{\varepsilon,P}$ for all $\varepsilon > 0$ and finite partitions P. On the other hand, let $\{P_i\}$ be a countable family of partitions labelled by finite subsets of $\mathbb{N}$, dense in the L^1-metric on partitions. As $h(T^\beta, P)$ is continuous in the L^1-metric, if $\beta \in \mathcal{O}_{1/k,P_i}$ for all k and i then we will have

$$h(T^\beta, P) \leq h_m(T^\alpha)$$

for all P and hence $h(T^\beta) = h_m(T^\alpha)$.

All we need to see is that the sets $\mathcal{O}_{\varepsilon,P}$ are open. This, though, is just upper semi-continuity of entropy $h(\tilde{\mu})$ for σ-invariant and ergodic measures on Σ_P^G, Theorem 3.0.17. □

Our goal now is to show that if ever $h_m(T^\alpha) < h(T^\alpha)$, for some α, then for this size m and for **any** G-arrangement β, we have $h_m(T^\beta) = 0$. Note that this β need not be in the m-equivalence class $E_m(\alpha)$ or even on the orbits of the same free and ergodic action. Although our path to this is somewhat technical we gain a lot of insight into the relation between a size and its entropy along the way.

First we show that if entropy can be lowered by moving to another m-equivalent arrangement, then relative to a partition, it can be lowered by an element of the full-group.

Lemma 5.0.5. *Let $e > 0$. Suppose, for some α,*

$$h_m(T^\alpha) < h(T^\alpha) - e.$$

Let P be any partition. For any $\varepsilon > 0$, there exists ϕ such that $m(\alpha, \phi) < \varepsilon$ and $h(T^{\alpha\phi}, P) < h(T^\alpha) - e$.

Proof As the full-group acts minimally on $E_m(\alpha)$ and preserves entropy, those $\beta \in E_m(\alpha)$ with

$$h(T^\beta) < h(T^\alpha) - e$$

are a dense set. Thus those β with

$$h(T^\beta, P) < h(T^\alpha) - e$$

are a dense and open set. On the other hand, those $\alpha\phi$ with

$$m(\alpha, \phi) < \varepsilon$$

are dense in $B_\varepsilon(\alpha)$, the ball of radius ε in m about α. Intersecting these two sets yields the result. □

Next we show that if entropy can be lowered, then it can be lowered on an independent and identically distributed (i.i.d.) process. (We are following the argument of [25] and [43] here. Knowing that the m-entropy of a Bernoulli action is less than its entropy will allow us to conclude this for all ergodic systems of positive entropy.)

Lemma 5.0.6. *Let α be any arrangement. Suppose there exists $e > 0$ such that $h_m(T^\alpha) < h(T^\alpha) - e$. Let (U^β, P) be any i.i.d. process with $h(U^\beta, P) = h(T^\alpha)$. For any $\varepsilon > 0$, there exists ϕ in the full-group of U^β such that $m(\beta, \phi) < \varepsilon$ and $h(U^{\beta\phi}, P) < h(U^\beta, P) - e$.*

Proof Fix $\alpha \in \mathscr{A}$ with $h(T^\alpha) > 0$. Suppose $e > 0$ such that

$$h_m(T^\alpha) < h(T^\alpha) - e.$$

Let (U^β, P) be any i.i.d. process with $h(U^\beta, P) = h(T^\alpha)$. Let $\varepsilon > 0$. Let $\eta > 0$, so that $h_m(T^\alpha) < h(T^\alpha) - e - \eta$.

At this point, as in both [25] and [43] we seem forced to introduce a circularity in that we want to apply Sinai's Theorem to say we can obtain a full-entropy i.i.d. factor of T^α. This is of course a known theorem, so no circularity is really obtained. We also note that even within the confines of our work here no circularity is obtained as for the size m of conjugacy, m-entropy is of course entropy and the Sinai's Theorem (our Theorem 7.3.3) for this strongest of equivalences will hold without further ado concerning m-entropy.

Applying the Ornstein–Weiss, Sinai Theorem [37] we see that there exists a partition P' such that $h(T^\alpha, P') = h(T^\alpha)$ and (T^α, P') has the same i.i.d. distribution as (U^β, P), that is to say

$$\|(T^\alpha, P'), (U^\beta, P)\|_* = 0.$$

By Lemma 5.0.5 for any $\varepsilon > 0$ there exists a full-group element ϕ' with $m(\alpha, \phi') < \varepsilon$ such that $m(\alpha, \phi') < \frac{\varepsilon}{2}$ and

$$h(T^{\alpha\phi'}, P') < h(T^\alpha) - e - \eta.$$

If, in fact, this ϕ' were measurable with respect to $\bigvee_{g \in G} T^\alpha_{g^{-1}}(P')$ then ϕ' would have a "version" on (U^β, P), which would complete the proof. In general this need not hold. We need to "copy" ϕ' into this process.

Without loss of generality, we may assume that ϕ' is bounded. Select $\delta_1 > 0$, $\delta_1 = \delta_1(\alpha, \phi', \varepsilon, \eta)$, from Axiom 3 and from Theorem 3.0.17.

Thus if

$$\|(T^\alpha, g_{(\alpha,\phi')} \vee P'), (T^\alpha, g_{(\alpha,\phi)} \vee P')\|_* < \delta_1$$

then both

$$m(\alpha, \phi) < m(\alpha, \phi') + \frac{\varepsilon}{2}$$

and

$$h(T^{\alpha\phi}, P') < h(T^{\alpha\phi'}) + \eta.$$

Apply Corollary 4.0.9 with "Q"$= P'$ and "ϕ"$= \phi'$ to select ϕ, measurable with respect to $\mathscr{H} = \bigvee_{g \in G} T^\alpha_{g^{-1}}(P')$ such that

$$\|(T^\alpha, g_{(\alpha,\phi')} \vee P'), (T^\alpha, g_{(\alpha,\phi)} \vee P')\|_* < \delta_1.$$

Hence

$$h(T^{\alpha\phi}, P') < h(T^\alpha) - e,$$

which completes the result. □

Next, we see that if the entropy of a process can be perturbed down by arbitrarily small (in m) elements of the full-group, then the m-entropy of the action is in fact less than the standard entropy. This provides a converse to Lemma 5.0.6

Lemma 5.0.7. *Let $e > 0$. Suppose, for some partition P and arrangement α, there exists $\{\phi_i\} \subseteq \Gamma$ with $m(\alpha, \phi_i) \to 0$ and*

$$h(T^{\alpha\phi_i}, P) < h(T^\alpha, P) - e$$

for all i. Then $h_m(T^\alpha) \leq h(T^\alpha) - e$.

Proof To begin, as the assumed inequality above is strict, it still holds for some $e' > e$. By Theorem 3.0.17 conditional entropy is upper semi-continuous with respect to the distribution pseudometric. Thus for any finite partition Q, we have that

$$\limsup_{i\to\infty} h(T^{\alpha\phi_i}, P \vee Q) \leq \limsup_{i\to\infty} (h(T^{\alpha\phi_i}, P) + h(T^{\alpha}, Q|P)).$$

Thus, by hypothesis, for any finite partition Q,

$$\limsup_{i\to\infty} h(T^{\alpha\phi_i}, P \vee Q) < h(T^{\alpha}, P \vee Q) - e',$$

a strict inequality. In particular this tells us that for any finite partition Q there is a partition $R = P \vee Q$ refining Q and elements of the full-group ϕ_i with $m(\alpha, \phi_i) \to 0$ and

$$\limsup_{i\to\infty} h(T^{\alpha\phi_i}, R) < h(T^{\alpha}, R) - e'.$$

Let P_j be a countable list of finite partitions dense in L^1 in the space of all partitions taking values in $\mathbb{N}$. We will define a sequence of sets. Let $\mathcal{O}_j$ consist of those $\beta \in E_m(\alpha)$ for which there is some partition R refining P_j and

$$h(T^{\beta}, R) < h(T^{\alpha}, R) - e'.$$

Upper semi-continuity of entropy tells us this is an open set. Using $\beta = \alpha\phi_i$ for i large enough we see that $\mathcal{O}_j$ has α as an accumulation point. Consider those ϕ in the full-group that have finite order. By Theorem 4.0.2 such $\alpha\phi$ with ϕ of finite order are m-dense in $E_m(\alpha)$. We show that any such $\alpha\phi$ is an accumulation point of $\mathcal{O}_j$ showing it to be dense. Fix such a ϕ and notice that any partition now is refined by a finite ϕ-invariant partition (by invariant here we mean one whose elements are permuted by ϕ). As $(T^{\alpha\phi}, R\phi)$ is identical in distribution to (T^{α}, R) for such a ϕ-invariant partition we must have

$$h(T^{\alpha\phi}, R) = h(T^{\alpha}, R).$$

As $T^{\alpha\phi}$ is conjugate to T^{α} for any such R there must be a further refinement R', which we can once more assume is ϕ-invariant, and full-group elements ψ_i with $m(\alpha\phi, \psi_i) \to 0$ and

$$\limsup_{i\to\infty} h(T^{\alpha\phi\psi_i}, R') < h(T^{\alpha\phi}, R') - e' = h(T^{\alpha}, R') - e'.$$

This implies that once i is large enough $\alpha\phi\psi_i \in \mathcal{O}_j$ and $\alpha\phi$ is an accumulation point of $\mathcal{O}_j$ and these open sets are dense.

We know $E_m(\alpha)$ is Polish in the m-topology and hence these sets have non-trivial intersection in $E_m(\alpha)$. For any β in this intersection, and for all partitions P, there will be a refinement R of P with

$$h(T^\beta, R) < h(T^\alpha, R) - e'.$$

It follows easily that

$$h(T^\beta) \leq h(T^\alpha) - e' < h(T^\alpha) - e.$$

□

Lemma 5.0.8. *Let α be any arrangement. Suppose there exists $e > 0$ such that $h_m(T^\alpha) < h(T^\alpha) - e$. Let β be any arrangement on any free and ergodic orbit relation for which $h(T^\beta) \geq h(T^\alpha)$. Then $h_m(T^\beta) < h(T^\beta) - e$.*

Proof Apply the Ornstein–Weiss Sinai Theorem [37] to obtain a partition P such that (T^β, P) is i.i.d. with $h(T^\beta, P) = h(T^\alpha)$. By Lemma 5.0.6, there exists a sequence $\{\phi_i\} \subseteq \Gamma$ such that $m(\beta, \phi_i) \to 0$ and, for every i,

$$h(T^{\beta\phi_i}, P) < h(T^\beta, P) - e.$$

By Lemma 5.0.7, $h_m(T^\beta) < h(T^\beta) - e$. □

The next result gives some insight into the relation between a size m, rearrangements and entropy. This is precisely where we use the copying lemma Theorem 4.0.13 by way of Corollary 4.0.14.

Theorem 5.0.9. *Suppose T^α is an ergodic G-action, P is a finite partition, $e > 0$. Suppose $\{\phi_i\} \subseteq \Gamma$ with*

(i) $m(\alpha, \phi_i) \underset{i}{\to} 0$, *and*

(ii) $h(T^\alpha, P | P \circ \phi_i^{-1}) > e$.

Then there exists $\{\psi_i\} \subseteq \Gamma$ such that

(iii) $m(\alpha, \psi_i) \underset{i}{\to} 0$, *and*

(iv) $h(T^\alpha, P \circ \psi_i^{-1}) = h(T^{\alpha\psi_i}, P) < h(T^\alpha, P) - e$.

Proof Given $\varepsilon > 0$, we will construct ψ with

$$m(\alpha, \psi) < \varepsilon \text{ and } h(T^{\alpha\psi}, P) < h(T^\alpha, P) - e.$$

Choose ϕ_i with $m(\alpha, \phi_i) < \varepsilon$. Axiom 3 tells us that there is a $\delta > 0$ so that if

$$\|(\alpha, \phi_i), (\alpha, \psi)\|_* < \delta$$

then we will still have $m(\alpha, \psi) < \varepsilon$. Corollary 4.0.14 tells us we can construct a ψ with

$$h(T^{\alpha\psi}, P) = h(T^\alpha, P \circ \psi^{-1}) < h(T^\alpha, P) - e$$

and with

$$\|(\alpha, \phi_i), (\alpha, \psi)\|_* < \delta$$

giving the conclusion. □

The preceding result can be read as saying that if one can find full-group elements of arbitrarily small m-size that move a partition in a way visible to entropy, then one can actually lower entropy and as we are about to see, push entropy to zero. This result will be central to the rest of our proof of the dichotomy of entropy-preserving and entropy-free sizes.

Lemma 5.0.10. *Let $e > 0$. Let T_1 and T_2 be two ergodic G-actions such that $h(T_1 \times T_2) > 0$ and $h_m(T_1 \times T_2) < h(T_1 \times T_2) - e$. Then either $h_m(T_1) < h(T_1) - \frac{e}{2}$ or $h_m(T_2) < h(T_2) - \frac{e}{2}$.*

Proof Let α_1 and α_2 be two G-arrangements with $T_1 = T^{\alpha_1}$ and $T_2 = T^{\alpha_2}$. Associated with $T^{\alpha_1} \times T^{\alpha_2}$ is a $G \times G$-arrangement α such that $T^\alpha = T^{\alpha_1} \times T^{\alpha_2}$.

Thus, according to the hypotheses, we have a $G \times G$-arrangement α such that $h(T^\alpha) > 0$ and $h_m(T^\alpha) < h(T^\alpha) - e$.

Apply Lemma 5.0.6 to see that for any i.i.d. process (U^β, P), if $h(U^\beta, P) = h(T^\alpha)$ then for any $\varepsilon > 0$, there exists ϕ in the full-group of U^β with $m(\beta, \phi) < \varepsilon$ and $h(U^{\beta\phi}, P) < h(U^\beta, P) - e$.

For $i = 1, 2$, let (U^{β_i}, P_i) be i.i.d. processes with $h(U^{\beta_i}, P_i) = h(T^{\alpha_i})$. Let $P = P_1 \times P_2$, β be such that $U^\beta = U^{\beta_1} \times U^{\beta_2}$. Then for the ϕ found above $m(\beta, \phi) < \varepsilon$ and $h(U^{\beta\phi}, P_1 \times P_2) < h(U^\beta, P_1 \times P_2) - e$.

Note that, in general, $U^{\beta\phi}$ is not a direct product. The above calculation together with basic entropy facts, imply that (writing 2 to mean the trivial algebra) one of the following must hold; either

(1) $h(U^\beta, P_1 \times 2 | (P_1 \times 2) \circ \phi^{-1}) > \frac{e}{2}$, or
(2) $h(U^\beta, 2 \times P_2 | (2 \times P_2) \circ \phi^{-1}) > \frac{e}{2}$.

Without loss of generality, suppose (1) holds. By Theorem 5.0.9, given any $\bar{\varepsilon} > 0$, we could have selected ε in such a way that there exists ψ in the full-group of U^β with $m(\beta, \psi) < \bar{\varepsilon}$ and

$$h(U^\beta, (P_1 \times 2) \circ \psi^{-1}) = h(U^{\beta\psi}, P_1 \times 2) < h(U^\beta, P_1 \times 2) - \frac{e}{2}.$$

Apply Corollary 4.0.9, with "Q"$= P_1 \times 2$ and

$$\mathscr{H} = \vee_{g \in G} U^\beta_{g^{-1}}(P_1 \times 2) = \vee_{g \in G} U_1^{\beta_1}{}_{g^{-1}}(P_1).$$

Thus for any $\delta > 0$ there exists $\tilde{\psi}$ in the full-group of $U_1^{\beta_1}$

$$\|(U^\beta, g_{(\beta,\psi)} \vee (P_1 \times 2)), (U_1^{\beta_1}, g_{(\beta,\tilde{\psi})} \vee (P_1))\|_* < \delta.$$

By Axiom 3, we may select δ so small that also $m(\beta_1, \tilde{\psi}) < \varepsilon$. Furthermore, by the upper semi-continuity of entropy, for any η we may select δ so small that $h(U_1^{\beta_1\tilde{\psi}}, P_1) < h(U^{\beta\psi}, P_1 \times 2) + \eta$. Thus, in fact, the δ may be selected so small that we still have

$$h(U_1^{\beta_1\tilde{\psi}}, P_1) < h(U^\beta, P_1 \times 2) - \frac{e}{2} = h(U_1^{\beta_1}, P_1) - \frac{e}{2}.$$

In particular, for any $\varepsilon > 0$, we have produced an element $\tilde{\psi}$ in the full-group of $U_1^{\beta_1}$ such that $m(\beta_1, \tilde{\psi}) < \varepsilon$ and $h(U_1^{\beta_1\tilde{\psi}}, P_1) < h(U_1^{\beta_1}, P_1) - \frac{e}{2}$.

Now, Lemma 5.0.7 implies that $h_m(T_1) < h(T_1) - \frac{e}{2}$ as T_1 must have a factor conjugate to the Bernoulli shift $U_1^{\beta_1}$. □

We are now ready to complete the proof that there are two kinds of sizes – those which are entropy preserving, in which case the m-entropy is simply entropy, and those which are entropy-free, in which case the m-entropy is always zero.

Theorem 5.0.11. *Let $e > 0$. Suppose there exists some G-arrangement α such that $h(T^\alpha) > 0$ and $h_m(T^\alpha) < h(T^\alpha) - e$. Then for any free and ergodic G-action S^β, $h_m(S^\beta) = 0$.*

Proof Suppose for some arrangement β, $h_m(T^\beta) = B > 0$. By Theorem 5.0.4, we may select $\gamma \in E_m(\beta)$ such that $h(S^\gamma) = h_m(S^\beta) = B > 0$. In particular, $h(S^\gamma) = h_m(S^\gamma) = B > 0$.

Let (U, P) be any Bernoulli process with $h(U, P) = B$. Select k so large (but finite!) that the k-fold product $(U \times \cdots \times U, P \times \cdots \times P)$ has entropy $h(U \times \cdots \times U, P \times \cdots \times P) \geq h(T^\alpha)$. By Lemma 5.0.8, $h_m(U \times \cdots \times U) < h(U \times \cdots \times U) - e$. Applying Lemma 5.0.10 inductively, $k-1$ times, we see that $h_m(U) < h(U) - \frac{e}{2^{k-1}}$. (In other words, we can lower the entropy on any Bernoulli process.) Since $h(S^\gamma) = h(U)$, Lemma 5.0.8

now implies that $h_m(S^\gamma) < h(S^\gamma) - \frac{e}{2^{k-1}}$. But this is a contradiction, since $h_m(S^\gamma) = h(S^\gamma)$. Hence, we must have that $h_m(S^\beta) = 0$ universally. □

This completes our technical work on m-entropy.

Definition 5.0.12. *Let m be a given size.*

(1) *If* $h_m(T^\alpha) = h(T^\alpha)$, *for all arrangements* α, *then* m *is called an* **entropy-preserving** *size.*
(2) *If* $h_m(T^\alpha) = 0$ *for all arrangements* α, *then* m *is called an* **entropy-free** *size.*

6
m-joinings

6.1 Polish topologies

Our objective here is to produce a topological model for arrangements and rearrangements. We seek a single action of G by homeomorphisms of a topological space on which there is a canonical notion of a base arrangement and a rearrangement of it given as continuous functions. We want this model to be sufficiently large that any example of a measure space, an arrangement and rearrangement will be conjugate to this model space for a particular choice of measure. Thus to move from one measurable example to another is to choose a different invariant measure on the model. This will allow us to topologize the set of all rearrangements by a weak*-topology on this space of measure. To accomplish this, the topological space must be rather elaborately defined, and in particular cannot be compact.

Sufficient to our purposes though they will all be Polish spaces in that the topological space is metrizable and, for some choice of metric, is complete and separable. Usually we will give explicit metrics for these spaces, and quite often they will not be complete for that metric, but will always sit as residual subsets of their completion, as this is a universal property independent of the choice of metric. We will rely in an essential way on the results of Varadarajan [57] that the space of probability measures on a Polish space is again a Polish space in the weak*-topology of convergence on bounded continuous functions.

Recall the following definition.

Definition 6.1.1 ([5]). *A topological space (X, τ) is called a* **Polish space** *if it can be metrized as a complete and separable metric space.*

Polish spaces are functorially quite robust. Of particular interest to us, are those listed in the following lemma.

Lemma 6.1.2.

(i) *Any countable set in the discrete topology is a Polish space.*
(ii) *Any countable product of Polish spaces (in the product topology) is a Polish space.*
(iii) *Any open, closed or G_δ subset of a Polish space is a Polish space.*

Proof See [5]. □

One fact about Polish spaces, which will be important for our work, is that a Polish space is a universal G_δ, that is to say, a G_δ-subset of its completion, relative to any metric giving the topology. In particular, we have the following lemma from [5].

Lemma 6.1.3. *Let X be a Polish space. Then a subspace of X is Polish if and only if it is a G_δ in X.*

Proof See [5]. □

Definition 6.1.4. *Let $\mathcal{M}(X)$ be the space of real valued Borel measures on the Polish space (X, τ) (we suppress the topology in the notation). Let $\mathcal{M}_1(X)$ be the probability measures in $\mathcal{M}(X)$.*

Theorem 6.1.5 ([57]). *In the weak*-topology of convergence on bounded continuous functions, $\mathcal{M}_1$ is again a Polish space.*

Proof Theorems 14 and 17 of [57] give this result. Note: "topologically complete and separable" is equivalent to Polish. □

We find the following picture useful to understand this result.

Lemma 6.1.6. *The space (X, τ) is Polish if and only if it is a residual subset of some compact metric completion.*

Proof If (X, τ) is a residual subset of some compact metric space, it is separable, and, as any G_δ subset of a complete metric space is topologically complete, we have one direction of the proof. For the other, suppose (X,d) is a complete and separable metric space. Let x_i be a countable and dense subset of X. Define a map

$$h(x) = \{d(x_i, x)\}_{i=1}^{\infty}.$$

It is easy to check that h is 1-1 and a homeomorphism to its image in $\mathbb{R}^{\mathbb{N}}$ with the product topology.

For each x_i and $j \geq 1$ define an open set $\mathcal{O}_{i,j} \subseteq \mathbb{R}^{\mathbb{N}}$ by

$$\mathcal{O}_{i,j} = \{\vec{y} : |d(x_k, x_i) - y_k| < 2^{-j} \text{ for all } k \leq \max(i,j)\},$$

that is to say, all vectors $\vec{y}$ which are within 2^{-j} of $h(x_i)$ on coordinates $1, 2, \ldots, \max(i,j)$.

For any $x \in X$ there is an i with $d(x, x_i) \leq 2^{-j}$ and hence for all k,

$$|d(x_k, x_i) - d(x_k, x)| < 2^{-j} \text{ and so } h(x) \in \mathcal{O}_{i,j}.$$

Thus $\mathcal{O}_j = \bigcup_i \mathcal{O}_{i,j}$ is an open cover for the range of h, and

$$\text{range}(h) \subseteq \bigcap_j \mathcal{O}_j.$$

Suppose $\vec{y} \in \bigcap_j \mathcal{O}_j$. Then in fact there must be a sequence $i(j)$ with

$$\vec{y} = \bigcap_j \mathcal{O}_{i(j),j}$$

(it is easy to see that such an intersection is at most a single point).

Thus in particular for any $j \geq 1$,

$$|d(x_{i(j)}, x_{i(j)}) - y_{i(j)}| = |y_{i(j)}| \leq 2^{-j}$$

and as we know for any $j' \geq \max(i(j), j)$,

$$|d(x_{i(j)}, x_{i(j')}) - y_{i(j)}| < 2^{-j'} \leq 2^{-j},$$
$$|d(x_{i(j)}, x_{i(j')})| < 2^{-j+1}.$$

This implies $x_{i(j)}$ is Cauchy in X, converging to some x_0.

For any $k \geq 0$, if $\sup(j, i(j)) \geq k$ then $|d(x_k, x_{i(j)}) - y_k| < 2^{-j}$. As $j \to \infty$, $d(x_k, x_{i(k)}) \to d(x_k, x)$ and so $h(x) = \vec{y}$ and $\text{range}(h) = \bigcap_j \mathcal{O}_j$ is a G_δ subset of $\mathbb{R}^{\mathbb{N}}$. Now $\mathbb{R}^{\mathbb{N}}$ can be compactified as a residual subset of $[0,1]^{\mathbb{N}}$ and hence (X,d) can be embedded as a G_δ subset of this compact metric space. □

Lemma 6.1.7. *If (X, τ) is a Polish space and $Y \subset X$ is a G_δ subset (and hence Polish) then $\mathcal{M}(Y)$ consists of the subset of $\mathcal{M}(X)$ of measures supported on Y with the inherited topology.*

Proof This is a direct consequence of Theorem 2.7, page 368 of [21], that if $m_i \underset{i}{\rightarrow} m$ in $\mathcal{M}(X)$ and g is bounded and continuous m-a.e., then

$$\int_X g\, d\mu_i \to \int_X g\, d\mu.$$

□

The above two lemmas tell us that we can view a Polish space as a residual subset of a compact metric space, and its space of Borel measures as just the subset of Borel measures on the compactification, with the inherited weak*-topology. Most importantly, this is independent of our choice of metric compactification.

To bring dynamics into the picture we say $(X, \tau, \{T_g\}_{g\in G})$ is a Polish G-action if (X, τ) is a Polish space and $\{T_g\}_{g\in G}$ is an action of G on X by homeomorphisms.

For example, if G is countable and (X, τ) is a Polish space, then the shift action of G on X^G is a Polish G-action. If G is countable notice that, for this example, there is a metric compactification of X to which the action of G extends as homeomorphisms (if (C, d) is some metric compactificication of X, then C^G is a metric compactification of X^G).

Lemma 6.1.8. *If $(X, \tau, \{T_g\}_{g\in G})$ is a Polish G-action, G a countable group, then X has a metric compactification to which $\{T_g\}_{g\in G}$ extends as homeomorphisms.*

Proof Embed X in X^G by mapping x to $\{T_g(x)\}_{g\in G}$. Call the map p. Note that p conjugates $\{T_g\}_{g\in T}$ to the shift action $\{\sigma_g\}_{g\in G}$. It is easy to see that p is a homeomorphism from X to its range. Let C be some metric compactification of X, and d some metric on C^G giving the product topology. The pull-back $d \circ p$ of d to X via p is a metric giving the topology τ. The completion of X with respect to this metric will embed via p isometrically to the closure of X in C^d, a compact metric space. As $p(X)$ is σ-invariant, so is its closure. □

For $(X, \tau, \{T_g\}_{g\in G})$ a Polish G-action, G a countable and discrete group, for any point $x \in X$, the set $I(x) = \{g \in G : T_g(x) = x\}$ is a subgroup. Similarly, for any $g \in G$, $\{x \in X : T_g(x) = x\}$ is a closed subset. For any subgroup $H \subseteq G$ we can define $X_H = \{x : H = I(x)\}$.

Lemma 6.1.9. *For any subgroup $H \subseteq G$, $\{x \in X : I(x) \subseteq H\}$ is a closed subset, and X_H is a G_δ subset of it. In particular each X_H is a Polish space.*

Proof Certainly $\{x \in X : I(x) \subseteq H\} = \bigcap_{g \in H}\{x \in X : T_g(x) = x\}$ is a closed set. Now

$$X_H = \{x \in X : I(x) \subseteq H\} \setminus \bigcup_{g \notin H} \{x \in X : T_g(x) = x\}$$

a closed set with a countable union of closed sets removed, hence a G_δ. □

Lemma 6.1.10. *Let $(X, \tau, \{T_g\}_{g \in G})$ be a Polish G-action. The subset of $\mathcal{M}_1(X)$ consisting of all $\{T_g\}_{g \in G}$-invariant measures is a (perhaps empty) closed and convex subset of $\mathcal{M}_1(X)$. The extreme measures are the ergodic measures for the action and every invariant measure is an integral of ergodic and invariant measures.*

Proof If (C, d) is a metric completion to which $\{T_g\}_{g \in G}$ extends homeomorphically, then in $\mathcal{M}_1(C)$ the invariant measures are a compact and convex subset. Hence the intersection with $\mathcal{M}_1(X)$ is relatively closed. As $\mathcal{M}_1(X)$ is a convex subset of $\mathcal{M}_1(C)$, we obtain convexity for the invariant measures.

If we write an invariant measure μ as an integral of extreme invariant measures in $\mathcal{M}_1(C)$ (as we know we can, this is just the ergodic decomposition)

$$\mu = \int \nu \, d\eta(\nu)$$

where η is a probability measure supported on the ergodic measures in $\mathcal{M}_1(C)$, it is an easy calculation that η-a.e. ν must be supported on X, that is to say must be in $\mathcal{M}_1(X)$ as μ is. Hence η must be supported on the extremal measures in $\mathcal{M}_1(X)$. It is clear that if μ is ergodic for the action on C it is ergodic for, its restriction to X is. □

This implies that the space of invariant probability measures for a Polish G-action is a Polish space. We let $\mathcal{M}_s(X)$ represent the invariant (stationary) measures, and $\mathcal{M}_e(X)$ the ergodic ones, suppressing the other terms of $(X, \tau, \{T_g\}_{g \in G})$ unless they are not obvious.

Corollary 6.1.11. *Suppose $(X, \tau, \{T_g\}_{g \in G})$ is a Polish G-action, G a countable and discrete amenable group. The space of ergodic and invariant measures $\mathcal{M}_e(X)$ in the weak* topology is a G_δ-subset of $\mathcal{M}_s(X)$ and hence is a Polish space.*

Proof It is sufficient to prove this for $(X,\tau) = (C,d)$ a compact metric space where it is a standard consequence of the mean ergodic theorem. □

The next definition and theorem are not precisely to our purpose, but indicate a bit of the impact of this point of view, and are a first case of the general machinery we are building.

Definition 6.1.12. *Suppose (X_1, τ_1, T_g^1) and (X_2, τ_2, T_g^2) are two Polish G-actions, G a countable and discrete amenable group. Suppose μ_1 and μ_2, respectively, are ergodic and invariant measures for these actions. The* **joinings** *of μ_1 and μ_2 will consist of all invariant measures for the Polish action $(X_1 \times X_2, \tau_1 \times \tau_2, \{T_g^1 \times T_g^2\}_{g \in G})$ which project to μ_1 and μ_2, respectively, on the two coordinates. We write this set of measures as $J(T_1, T_2)$ suppressing all the space variables.*

Theorem 6.1.13. *The space $J(T_1, T_2)$ is a closed, convex subset of $\mathcal{M}_s(X_1 \times X_2)$. Any joining $\hat{\mu} \in J(T_1, T_2)$ will be an integral of ergodic measures in $\mathcal{M}_e(X_1 \times X_2)$ which must themselves be joinings. The ergodic joinings themselves are a G_δ subset of $\mathcal{M}_1(X_1 \times X_2)$ and hence are a Polish space.*

Proof As in earlier arguments, we can again assume $X_1 = C_1$ and $X_2 = C_2$ are compact spaces by extending the actions to some metric compactification. In the context of compact spaces, the fact that any joining is an integral of ergodic joinings is well known (see for example [44]). As these ergodic joinings project to μ_1 and μ_2 they are supported on $X_1 \times X_2$. As $J(T_1, T_2)$ is a closed subset of $\mathcal{M}_1(C_1 \times C_2)$, and the ergodic joinings are the intersection $J(T_1, T_2) \cap \mathcal{M}_e(X_1 \times X_2)$, an intersection of a closed set and a G_δ, the ergodic joinings form a G_δ. □

We are now ready to introduce the explicit Polish G-actions we will use to model arrangements and rearrangements, and ultimately to define the notion of m-joinings for m a size. In this section we will construct a natural Polish space modeling a pair α and β of G-arrangements, that is to say a Polish topological space on which all such pairs can be modeled by invariant measures for a fixed orbit-equivalent pair of G-actions.

This is a dramatic shift in perspective. It is analogous to shifting from the consideration of all Σ-valued partitions of some fixed space and G-action to the space of all σ-invariant measures on Σ^G. In the latter there is only one partition of real interest, the time-identity partition.

Analogously in our case there will be only a distinctly limited and precisely encoded set of arrangements and rearrangements of interest on the spaces constructed. These will not be arrangements and rearrangements in the precise sense we have given earlier, as they will not sit on a single measure space. Rather, they will be topological analogues. There will always be a base arrangment which we will write as $\boldsymbol{\alpha}$ corresponding to the base action of G defined on the space. This basic action will be some generalized "shift" action which will be called $\sigma = \sigma^{\boldsymbol{\alpha}}$ if it is truly a shift on a product space, or $S = S^{\boldsymbol{\alpha}}$ otherwise. There will also always be some second arrangement (we are modeling changes of arrangements after all) which we will label $\boldsymbol{\beta}$ with corresponding actions $\sigma^{\boldsymbol{\beta}}$ or $S^{\boldsymbol{\beta}}$.

At times we will define the actions $S^{\boldsymbol{\alpha}}$ and $S^{\boldsymbol{\beta}}$ as maps before discovering that they are orbit-equivalent. Furthermore, for most of our work no measure will be specified. In the end though, two such actions will be orbit-equivalent, but may not be free. Hence the actions cannot be said to give rise to arrangements $\boldsymbol{\alpha}$ and $\boldsymbol{\beta}$. At such points we ask the reader to regard this notation as formal and to look for later clarification of what precisely $\boldsymbol{\alpha}$ and $\boldsymbol{\beta}$ represent.

Later we will also consider full-group elements intrinsic to the construction. In these models there will first be just one full-group element which we will write as $\boldsymbol{\phi}$, and later a sequence of them written $\boldsymbol{\phi}_i$. Whenever we use this bold-face notation we are indicating the canonical elements of the construction and not an arbitrary element of a full-group.

Our topological models will contain all pairs of arrangements, and later all rearrangements, and convergent sequences of rearrangements, not as distinct arrangements on one G-action, but as distinct invariant measures on one canonical construction.

As usual G is a fixed, countable and discrete amenable group, and F_i some Følner sequence in it.

We will stick to certain notational conventions that should help the reader follow some of the more technical twists we take. For a G-action $(Z, \mathscr{F}, \mu, T^{\alpha})$, the symbol p will always represent the function from a point z to its name in Z^G, i.e. $p(z) = \{T_g^{\alpha}(z)\}_{g \in G}$. The symbol q will always represent a standard involution of the spaces involved. The symbols π (π_1 or π_2) will always represent the projection of a product onto one of its coordinates, i.e. $Z_1 \times Z_2$ onto Z_1 or Z_2. We will develop more elaborate topological representations of a G-action than simply the shift map on Z^G. Under these circumstances we will us L to represent a labeling of the action by some more elaborate coordinates.

Remember that to any two arrangements α and β and $x \in X$ we can associate a bijection $h_x^{\alpha,\beta} : G \to G$ which describes the manner in which β

arranges an α orbit. Similarly, to any rearrangement pair (α, ϕ) and point $x \in X$ we associate a map $f_x^{\alpha,\phi}$ which describes how ϕ moves points on the orbit of x as seen by the α-arrangement of that orbit. The actions T^α and T^β (or $T^{\alpha\phi}$) act on the spaces of such functions in an equivariant fashion by acting on the point x. It is these actions on the corresponding functions that we wish to model topologically. Keeping this in mind will make our definitons and calculations more transparent.

6.1.1 Overview of the topology on m-joinings

Before beginning the explicit construction of these models we give a brief and vague description of each to give an overview of their structure:

1. We start with $\mathcal{G}$, the space of all bijections of G fixing the identity, topologized as a complete metric space. This gives a symbolic model for a pair of rearrangements $\boldsymbol{\alpha}$ and $\boldsymbol{\beta}$ on the same orbits.

2. We take $\mathcal{M}_e(\mathcal{G})$ to be the ergodic measures for these arrangements with the weak*-topology.

3. Here, and generally to eliminate a lack of freeness of the symbolic actions, we extend the symbolic model by a coordinate of the form Z^G with the shift action, where Z is some non-trivial Polish space and call the enlarged space Y. We set $\mathcal{M}_e(Y)$ to be the invariant and ergodic measures on Y.

4. We next construct a symbolic model $\mathcal{R}$ for the action of an element $\boldsymbol{\phi}$ of the full-group on an orbit as a bijection of G and in $Y \times \mathcal{R}$ we restrict to the subset where $\boldsymbol{\beta} = \boldsymbol{\alpha\phi}$ giving a symbolic model for a rearrangement pair $(\boldsymbol{\alpha}, \boldsymbol{\phi})$.

5. To model a convergent sequence of rearrangements we take $\mathcal{R}^{\mathbb{N}}$ and restrict to those sequences for which the image rearrangements $\boldsymbol{\alpha\phi}_i$ converge in a particularly strong fashion. This set is written as the lim sup of a sequence of closed sets F_I in $\mathcal{R}^{\mathbb{N}}$ which is not a Polish space.

6. We now make a critical shift in perspective to a space $\mathcal{M}_*$ of measures ν on $\mathcal{R}^{\mathbb{N}}$ relative to which the $\boldsymbol{\alpha\phi}_i$ converge rapidly in $L^1(\nu)$. This space of measures is shown to be weak* Polish and supported on $\limsup_I F_I$.

7. We now define the space $\mathcal{M}_{**}$ to consist of those $\nu \in \mathcal{M}_*$ for which the canonical involution from $\boldsymbol{\alpha\phi}_i$ to $\boldsymbol{\beta\phi}_i^{-1}$ yields an image measure still in $\mathcal{M}_*$. We can once more add on a Z^G coordinate to create $\mathcal{M}_{**}(Z)$.

8. To add a size m to the picture we restrict the measures in $\mathcal{M}_{**}(Z)$ to those measures for which the corresponding sequences $\boldsymbol{\alpha\phi}_i$ and $\boldsymbol{\beta\phi}_i^{-1}$ are m-Cauchy at a uniform rate. This new space, called $\mathcal{M}^m(Z)$ is a weak*

Polish space of measures and models all m-equivalences via explicit choice of a rapidly converging sequence of rearrangements.

9. To build a space of m-joinings of two actions $T_1^{\alpha_1}$ and $T_2^{\alpha_2}$ we first model them as acting as the shift map on spaces Z_1^G and Z_2^G. We then consider the subset of $\mathcal{M}^m(Z_1 \times Z_2)$ where the action given by $\boldsymbol{\alpha}$ projected to Z_1^G is $T_1^{\alpha_1}$ and the action given by $\boldsymbol{\beta}$ projected to Z_2^G is $T_2^{\alpha_2}$. We call this Polish space of measures $J_m(T_1^{\alpha_1}, T_2^{\alpha_2})$.

Now to the actual construction of the models.

6.2 Modeling pairs of arrangements

Let $\mathcal{G}$ be the set of all bijections of G fixing the identity, that is to say, the collection of all maps $h : G \to G$ that are 1-1 and onto and with $h(\mathrm{id}) = \mathrm{id}$. These form a group under composition. We can topologize $\mathcal{G}$ as a subset of G^G, putting on G^G the product topology as a product of discrete spaces. It is an exercise that the group action of composition is jointly continuous, and the map $q : h \to h^{-1}$ is a homeomorphism of $\mathcal{G}$. This is our initial version of the involution q.

In Section 2.1 we gave an explicit metric d for the space $\mathcal{G}$ making it a complete metric space. Recall that this metric was invariant for the involution $q : h \to h^{-1}$. It is useful to notice that $\mathcal{G}$ is a zero-dimensional space, and so in $\mathcal{G} \times \mathcal{G}$ those pairs (h_1, h_2) with $d(h_1, h_2) \leq d$ for any d form a clopen set.

Also keep in mind the basic description of the topology as the weakest topology for which all the functions $g : h \to h(g)$ are continuous. This is a separable topology (for each g_i there are only countably many choices for $h(g_i)$), and hence $\mathcal{G}$ is a Polish space.

There are two natural actions of G on $\mathcal{G}$. We will call them $S^{\boldsymbol{\alpha}}$ and $S^{\boldsymbol{\beta}}$, even though we will not see the arrangements for a while, and in fact as the actions are not free and we have specified no invariant measures, we cannot formally call them arrangements. They are given by

$$S_g^{\boldsymbol{\alpha}}(h)(k) = h(kg)h(g)^{-1}$$

and

$$S_g^{\boldsymbol{\beta}}(h) = S_{h^{-1}(g)}^{\boldsymbol{\alpha}}(h)$$

or

$$S_g^{\boldsymbol{\beta}}(h)(k) = h(kh^{-1}(g))g^{-1}.$$

Theorem 6.2.1. *The collections of maps* $\{S_g^\alpha\}_{g\in G}$ *and* $\{S_g^\beta\}_{g\in G}$ *are both G-actions on* $\mathscr{G}$ *by homeomorphisms. As* $S_g^\alpha(h) = S_{h(g)}^\beta(h)$, *they are orbit-equivalent. They also are conjugate as the map* q *conjugates* S_g^α *to* S_g^β.

Proof The following two calculations explain why we have G-actions:

$$\begin{aligned}
S_{g_1}^\alpha \circ S_{g_2}^\alpha(h)(k) &= S_{g_1}^\alpha\left(S_{g_2}^\alpha(h)\right)(k)\\
&= S_{g_2}^\alpha(h)(kg_1)\left(S_{g_2}^\alpha(h)(g_1)\right)^{-1}\\
&= h(kg_1g_2)\left(h(g_2)\right)^{-1}\left(h(g_1g_2)h(g_2)^{-1}\right)^{-1}\\
&= h(kg_1g_2)h(g_1g_2)^{-1} = S_{g_1g_2}^\alpha(h)(k)
\end{aligned}$$

and for S^β showing the conjugation of S_g^α to S_g^β will suffice:

$$\begin{aligned}
q(S_g^\alpha(h))(k) = (S_g^\alpha(h))^{-1}(k) &= h^{-1}(kh(g))g^{-1}\\
&= S_g^\beta(h^{-1})(k) = S_g^\beta(q(h))(k).
\end{aligned}$$

As G is a group this implies that all the maps S_g^α and S_g^β are 1-1. As the topology is the minimal one making all $h \to h(k)$ continuous, to verify continuity of S_g^α is just to verify continuity of $h \to S_g^\alpha(h)(k) = h(kg)h(g)^{-1}$, which has the same level sets as $h \to h(kg)$ which is continuous. That S_g^β is continuous is the same argument. □

It is evident that the action of S_g^α is not free (in particular, the identity bijection is a fixed-point of the G-action). Hence we cannot speak of arrangements associated with the two actions in the form of maps from the orbit relation to G. We almost can though up to the isotropy subgroups of points, and we do this by defining two *almost*-arrangements by

$$\alpha(h, S_g^\alpha(h)) = gI_1(h) \qquad \text{and} \qquad \beta(h, S_g^\beta(h)) = gI_2(h)$$

where $I_1(h)$ is the isotropy subgroup of elements of the S^α-action fixing h, and $I_2(h)$ is the subgroup of the S^β-action fixing h. It is easy to see these are well-defined on pairs of functions on the same orbit.

Notice that $S_{gI_1(h)}^\alpha(h) = S_g^\alpha(h)$ and so we can write:

$$\beta(h, S_{gI_1(h)}^\alpha(h)) = \beta(h, S_{h(g)}^\beta(h)) = h(g)I_2(h).$$

If we regard the orbit of h as identified bijectively by

$$S_g^\alpha(h) \to gI_1(h) \in G/I_1(h),$$

and similarly identified bijectively by

$$S_g^\beta(h) \to gI_2(h) \in G/I_2(h),$$

then the map which carries this one identification to the other is the map $gI_1(h) \to h(g)I_2(h)$.

Lemma 6.2.2. *The bijection h is a group isomorphism from $I_1(h)$ to $I_2(h)$ and moreover, for any $g \in I_1(h)$ and $k \in G$, $h(kg) = h(k)h(g)$.*

Proof Notice $g \in I_1(h)$ iff $h(k) = h(kg)h(g)^{-1}$ for all k. This may be rewritten as $h(kg) = h(k)h(g)$. We have already seen that $h(gI_1(h)) = h(g)I_2(h)$ and this completes the result. □

Thus h is well defined as a map from $G/I_1(h) \to G/I_2(h)$. As such, h is precisely the map which rearranges the $S^{\boldsymbol{\alpha}}$ and $S^{\boldsymbol{\beta}}$ orbits, identified on these spaces by the two actions. We should point out here that even though on the orbits of $S^{\boldsymbol{\alpha}}$ we may only have an almost-arrangment, that is to say a map on $G/I_1(h)$, the value h itself tells us how to rearrange an entire copy of G, not just the quotient space. This observation was in fact the origin of the construction. We wanted an action which modeled topologically all possible ways that a second arrangement arranges the orbits as viewed from a first arrangement.

Our goal now is to produce a topological dynamical system on which the invariant Borel measures model all possible pairs of arrangements on any given measurable action. Notice that we are close to having just that. If $(X, \mathcal{F}, \mu)$ is a standard probability space, with an ergodic measure-preserving orbit relation $\mathcal{O}$, and two G-arrangements α and β on it, then the map

$$L : x \to h_x^{\alpha,\beta} \qquad \text{maps} \qquad X \to \mathcal{G}$$

with the arrangement α identified with the *almost*-arrangement $\boldsymbol{\alpha}$, and similarly β with $\boldsymbol{\beta}$. That is to say, L conjugates T^α to $S^{\boldsymbol{\alpha}}$ and T^β to $S^{\boldsymbol{\beta}}$. Furthermore $L^*(\mu) \in \mathcal{M}_e(\mathcal{G})$. Thus the weak*-topology on $\mathcal{M}_e(\mathcal{G})$ can be used to topologize spaces with pairs of arrangements. This is an inadequate structure so far because of a lack of 1-1-ness and of course the lack of freeness of the orbits of $S^{\boldsymbol{\alpha}}$. For example, if $\alpha = \beta$ then $L^*(\mu)$ will be a point mass on the identity. Even if $L^*(\mu)$ is supported on free orbits, the map L^* will be far from unique, as it depends only on the way the orbits are rearranged and this may very well only generate a subalgebra of $\mathcal{F}$.

To remedy this, let Z be a Polish topological space and let σ be the shift G-action on Z^G. For any $h \in \mathcal{G}$ define a map $r_h : Z^G \to Z^G$ by

$$r_h(\vec{z})(g) = \vec{z}(h^{-1}(g)).$$

That is,

$$\sigma_g(r_h(\vec{z})) = r_{S^\beta_g(h)}\sigma_{h^{-1}(g)}(\vec{z})$$

which implies that the map

$$\hat{q} : (h, \vec{z}) \to (h^{-1}, r_h(\vec{z}))$$

conjugates the action of $S^\alpha_g \times \sigma_g$ to that of $S^\beta_g \times \sigma_{h^{-1}(g)}$.

Suppose $(X, \hat{T}^\alpha)$ is a free G-action by homeomorphisms of a Polish space X, and $L : X \to \mathcal{G}$ is a continuous map factoring $\hat{T}^\alpha$ onto S^α. Then the action of S^β on $\mathcal{G}$ lifts canonically to X as the action

$$\hat{T}^\beta_g(x) = \hat{T}^\alpha_{L(x)^{-1}(g)}(x).$$

Notice that, in this case,

$$h^{\alpha,\beta}_x = L(x) \in \mathcal{G}.$$

We hope the reader will understand our continuing use of the notation α and β even though the space has changed.

The simplest case of this situation is where σ is the G-shift on Z^G, Z some Polish space and $\hat{T}^\alpha = S^\alpha \times \sigma$ on some Polish $Y \subseteq \mathcal{G} \times Z^G$, projecting to S^α on the first coordinate. By Lemma 6.1.9 the set of all free orbits in $\mathcal{G} \times Z^G$ is such a Y.

Lemma 6.2.3. *The maps $\hat{T}^\beta_g$ are homeomorphisms of Y forming a Polish G-action with the same orbits as $\hat{T}^\alpha$.*

Proof The only issue here is to see that the $\hat{T}^\beta_g$ are homeomorphisms and map Y to Y. Just notice that if h and $h' \in \mathcal{G}$ are close enough, then $h(g) = h'(g)$. Thus if $\vec{z}$ and $\vec{z}' \in Z^G$ are close enough $\hat{T}^\alpha_{h(g)}(y)$ and $\hat{T}^\alpha_{h'(g)}(y')$ will be close. We already know that $(\hat{T}^\beta_g)^{-1} = \hat{T}^\beta_{g^{-1}}$ and they are invertible. As $\hat{T}^\beta$ preserves $\hat{T}^\alpha$ orbits, it must map Y to Y. □

This justifies our use in this context of the two names $\hat{T}^\alpha$ and $\hat{T}^\beta$ as they are orbit-equivalent, and we can now take α and β as the two corresponding arrangements. As Y consists of free orbits these are actually arrangements, short of the fact that arrangements were originally defined only on measure spaces, and here we have no specified measure.

We have already seen the calculation that implies

$$h^{\alpha,\beta}_{h,\vec{z}} = h \qquad \text{and} \qquad h^{\beta,\alpha}_{h,\vec{z}} = h^{-1}.$$

Recall that the map $\hat{q} : (h, \vec{z}) \to (h^{-1}, r_h(\vec{z}))$ is an involution conjugating $\hat{T}^{\alpha}$ to $\hat{T}^{\beta}$.

Notice that the invariant (and ergodic) measures for $\hat{T}^{\alpha}$ and $\hat{T}^{\beta}$ are identical. Hence $\hat{q}^*$ acts as an involution on both $\mathcal{M}_s(Y)$ and $\mathcal{M}_e(Y)$.

A standard probability space is, by definition, a Borel measure on a compact metrizable space. Hence any free G-action on such a space $(Z, \mathcal{B}, \mu, \{T_g\}_{g\in G})$ can be mapped to Z^G by $p : z \to \{T_g(z)\}_{g\in G}$, conjugating the action of T to the shift map σ. This map will carry μ over to a measure supported on the Polish space of free σ orbits. If we have two arrangements of these orbits by α and β (and we may as well assume $T = T^{\alpha}$), then we get a map $\hat{L} : Z \to \mathcal{G} \times Z^G$ given by

$$\hat{L} : z \to (h_z^{\alpha,\beta}, p(z)).$$

This map conjugates T^{α} to $\hat{T}^{\alpha}$ and T^{β} to $\hat{T}^{\beta}$. It is supported on free orbits, i.e. maps to the Polish space Y. This map is 1-1 and so carries μ to a measure in $\mathcal{M}_e(\hat{T})$ faithfully modeling the original system. If we were to interchange the roles of α and β in this construction, the new image measure would simply be $\hat{q}^*$ of the original one.

In this construction the choice of compact metric space Z is arbitrary as any two non-atomic standard probability spaces are, up to null-sets, the same. What it provides us with is a fixed space on which to model pairs of arrangements topologically.

Suppose we have two free and ergodic G-actions

$$(Z_1, \mathcal{B}, \mu_1, \{T_g\}_{g\in G}) \quad \text{and} \quad (Z_2, \mathcal{B}, \mu_2, \{S_g\}_{g\in G}).$$

We can define the notion of an **orbit joining** of these two. Let p_1 and p_2 be the maps from Z_1 and Z_2 to their orbit names in Z_1^G and Z_2^G.

Definition 6.2.4. *Consider the Polish space $\mathcal{G} \times (Z_1 \times Z_2)^G$ and construct on it, as above, the two actions $\hat{T}^{\alpha}$ and $\hat{T}^{\beta}$ corresponding to the arrangements α and β. Let Y be its space of free orbits and π_i be the projection of Y to Z_i^G, $i = 1, 2$. Any measure $\hat{\mu} \in \mathcal{M}_e(\hat{T})$ with the property that*

$$\pi_1^*(\hat{\mu}) = p_1^*(\mu_1)$$

and

$$\pi_2^*(\hat{q}^*(\hat{\mu})) = p_2^*(\mu_2)$$

is called an orbit joining of the two free processes.

Notice that if we reverse the order of the two processes, then the interchange of the Z_1 and Z_2 coordinates in Y will take an orbit joining in the Z_1, Z_2 order to one in the Z_2, Z_1 order.

It is worthwhile to note that the space of orbit-joinings is not empty as it contains $\delta_{\mathrm{id}} \times \mu_1 \times \mu_2$, corresponding to the usual notion of joining and $\boldsymbol{\alpha}$ will equal $\boldsymbol{\beta}$, $\hat{\mu}$-a.s. More generally any joining of the two actions can be extended to an orbit joining by appending a point mass on $\mathrm{id} \in \mathcal{G}$.

6.3 Modeling rearrangements

The two notions of a joining and an orbit joining of two G-actions are the extreme cases of what we really need, a restricted orbit joining. To define this we must bring sizes into the structure, and to do this we must bring into the model the behavior of elements of the full-group. This follows a path quite analogous to the development of $\mathcal{G}$. For $f \in G^G$ we have already defined H by $H(f)(g) = f(g)gf(\mathrm{id})^{-1}$. Notice that $H(f)$ is a map $G \to G$ fixing the identity. In all the following calculations it is important to remember that $^{-1}$ placed by a bijection of G is the inverse bijection, and by an element of G is the inverse element. This is particularly regretable in regards to $H(f)^{-1}(g)$ and $H(f)(g)^{-1}$ which have distinct meanings.

Notice that for $f_h(g) = h(g)g^{-1}$, $H(f_h) = h$ and hence H is onto.

Definition 6.3.1. *Let $\mathcal{R} = \{f \in G^G : H(f) \in \mathcal{G}\}$, that is to say, the set of maps $f : G \to G$ with $g \to f(g)gf(\mathrm{id})^{-1}$ a bijection. Topologize $\mathcal{R}$ with the product topology on G^G.*

Notice that the map H from $\mathcal{R}$ into $\mathcal{G}$ is continuous. (In fact the map H is continuous as a map $G^G \to G^G$.) Also notice that if $f \in \mathcal{R}$, then

$$\begin{aligned} S_g^{\alpha}(H(f))(k) &= H(f)(kg)H(f)(g)^{-1} \\ &= f(kg)kg(f(\mathrm{id}))^{-1}(f(g)g(f(\mathrm{id}))^{-1})^{-1} \\ &= f(kg)f(g)^{-1} = H(\sigma_g(f))(k). \end{aligned}$$

That is to say, the map H conjugates the action of σ on G^G to the action of S^{α} on $\mathcal{G}$. As dynamics on $\mathcal{R}$ we take the restriction of σ to this σ-invariant subset. The pull-back f_h of h to $\mathcal{R}$ described above is not equivariant. If it were, all arrangements would be cohomologous.

Lemma 6.3.2. *$\mathcal{R}$ is a G_δ subset of G^G.*

Proof We already know that $\mathcal{G}$ is a Polish space, hence is a G_δ subset of G^G. As $\mathcal{R}$ is the pull-back by a continuous map (H) of this G_δ, it also is a G_δ subset of G^G. □

In $\mathcal{G} \times \mathcal{R}$ we can restrict to the $\hat{T}^\alpha = S^\alpha \times \sigma$ invariant subset of all $(H(f), f)$. We will refer to this as the graph of H even though the coordinates are interchanged. As H is continuous, this set is homeomorphic to $\mathcal{R}$. For $f \in \mathcal{R}$, we can define a map on the graph of H by

$$\phi(S^\alpha_g(H(f)), \sigma_g(f)) = S^\alpha_{f(g)g}(H(f), \sigma_{f(g)g}(f))$$

which is to say,

$$\alpha(\hat{T}_g(H(f), f), \phi(\hat{T}_g(H(f), f))) = f(g) = \sigma_g(f)(\mathrm{id}).$$

That is, $\phi(\hat{T}_g(H(f), f)) = \phi(H(\sigma_g(f)), \sigma_g(f))$ and is well defined on the graph of H, even on non-free orbits. The following simple lemma explains how $\mathcal{R}$ models rearrangements.

Lemma 6.3.3. *Restricted to the graph of H, which is $\hat{T}^\alpha$-invariant, the almost-arrangements α and β satisfy*

$$\beta = \alpha\phi.$$

As the graph of H is homeomorphic to $\mathcal{R}$, the same identity holds on $\mathcal{R}$ itself.

Proof This is just the following calculation:

$$\begin{aligned}\beta(\hat{T}^\alpha_g(H(f), f), \hat{T}^\alpha_k(H(f), f)) &= H(f)(k)(H(f)(g))^{-1} \\ &= f(k)kg^{-1}f(g)^{-1} = \alpha(\phi(S^\alpha_g(H(f))), \phi(S^\alpha_k(H(f)))).\end{aligned}$$

□

On the graph of H as a subset of $\mathcal{G} \times G^G$, we can act by the orbit equivalent action $\hat{T}^\beta$, which as we have seen is simply $\hat{T}^{\alpha\phi}$. This is a special case of our earlier discussion, just adding on another coordinate to the actions S^α and S^β. In particular the action $\hat{T}^\beta$ is conjugated by $\hat{q}$ to the action of $\hat{T}^\alpha$.

The map $\hat{q}$ is defined globally on $\mathcal{G} \times G^G$ and it is easy to see that it does not preserve the graph of H. A simple recoding will remedy this. For $f \in G^G$, define $I(f)(g) = f(g)^{-1}$, i.e. just replace each value in the G-name f with its inverse. Notice that for any $f \in \mathcal{G}$, r_f and I commute, as r_f rearranges coordinates and I just flips values at each coordinate to their inverses. Thus letting

$$\hat{I}(h,\vec{z}) = (h, I(\vec{z})),$$

$\hat{I}$ commutes with $\hat{T}^\alpha$, $\hat{T}^\beta$ and $\hat{q}$.

Lemma 6.3.4. *On the graph of H in $\mathcal{G} \times \mathcal{R}$, the map $\hat{q}\hat{I}$ conjugates $\hat{T}^\beta$ to $\hat{T}^\alpha$, i.e. $\hat{q}\hat{I}\hat{T}^\beta = \hat{T}^\alpha\hat{q}\hat{I}$.*

Proof The only thing to check is that $\hat{q}\hat{I}$ preserves the graph of H. For $f \in \mathcal{R}$ set

$$\hat{f}(g) = f(H(f)^{-1}(g))^{-1} = (r_{H(f)}(f)(g))^{-1}$$

which is to say

$$\hat{f} = I(f \circ (H(f))^{-1}) = I(r_{H(f)}(f)).$$

We can calculate:

$$H(\hat{f}) \circ (H(f)(g)) = f(g)^{-1}H(f)(g)f(\mathrm{id}) = g,$$

and $H(\hat{f}) = (H(f))^{-1}$. This calculation also shows that

$$\hat{f}(g) = H(f)^{-1}(g)f(\mathrm{id})^{-1}g^{-1},$$

as H of the right-hand-side is $H(f)^{-1}$ and the right-hand side agrees with $\hat{f}$ at the identity. (Two f's with the same image under H that agree at the identity must be the same.)

Now notice

$$\hat{q}\hat{I}(H(f), f) = \hat{I}(H(\hat{f}), r_{H(f)}(f)) = (H(\hat{f}), \hat{f}).$$

Thus $\hat{q}\hat{I}$ preserves the graph of H and hence conjugates $\hat{T}^\beta$ to $\hat{T}^\alpha$ on this Polish space. It is an easy calculation to show that $\hat{\hat{f}} = f$ making the map $f \to \hat{f}$ a continuous involution conjugating $\sigma = \sigma^\alpha$ to σ^β. □

Notice in particular that we now know

$$\widehat{\sigma_g(f)} = \sigma_{H(f)^{-1}(g)}(\hat{f}),$$

or equivalently that

$$\sigma_g(\hat{f}) = \widehat{\sigma_{H(f)(g)}(f)}.$$

Lemma 6.3.5. *For any $f \in \mathscr{R}$ there is a unique solution $z = z(f)$ to the equation $f(z)z = \mathrm{id}$. It is*

$$z(f) = \hat{f}(\hat{f}(\mathrm{id})) = H(\hat{f})(\hat{f}(\mathrm{id})) = H(f)^{-1}(\hat{f}(\mathrm{id})).$$

This value now satisfies

$$\sigma_{z(f)}(f) = \phi^{-1}(f).$$

More generally,

$$z(\sigma_g(f)) = S_g^{\alpha}(H(f))^{-1}(\hat{f}(H(f)^{-1}(g))) = S_g^{\alpha}(H(\hat{f}))(\hat{f}(H(\hat{f})(g))).$$

Proof We know $H(f)(g) = f(g)gf(\mathrm{id})^{-1}$ is 1-1 and onto. Hence there is a unique z with $H(f)(z) = f(\mathrm{id})^{-1}$. This is the z we want. Notice that $\phi(f) = \sigma_{f(\mathrm{id})}(f)$ so $\phi(f)(g) = f(gf(\mathrm{id}))$. Letting

$$\phi(f)(g) = f'(g) = f(gf(\mathrm{id})),$$

we calculate that

$$z(f')^{-1} = f'(z(f')) = f(z(f')f(\mathrm{id})).$$

Notice that $z = f(\mathrm{id})^{-1}$ satisfies $z^{-1} = f(zf(\mathrm{id}))$ and as this has a unique solution, $z(f') = f(\mathrm{id})^{-1}$. Now we see:

$$\sigma_{z(\phi(f))}(\phi(f)) = \sigma_{z(f')}(\sigma_{f(\mathrm{id})}(f)) = \sigma_{f(\mathrm{id})}^{-1}\sigma_{f(\mathrm{id})}(f) = f.$$

To calculate z more explicitly, notice that $H(f)(f(\mathrm{id})) = f(f(\mathrm{id}))$ and $\hat{f}(H(f)(f(\mathrm{id}))) = f(f(\mathrm{id}))^{-1}$ and so $z(\hat{f}) = f(f(\mathrm{id}))$. But then symmetrically $z(f) = \hat{f}(\hat{f}(\mathrm{id}))$.

Using the conjugation $H(\sigma_g(f)) = T_g(H(f))$ we conclude

$$\begin{aligned} z(\sigma_g(f)) &= H(\sigma_g(f))^{-1}(\widehat{\sigma_g(f)}(\mathrm{id})) \\ &= T_g(H(f))^{-1}(\sigma_{H(f)^{-1}(g)}(\hat{f})(\mathrm{id})) \\ &= T_g(H(f))^{-1}(\hat{f}(H(\hat{f})(g))). \end{aligned}$$

Notice that this gives the natural symmetry:

$$H(\hat{f})(\hat{f}(\mathrm{id})) = z(f) \quad \text{and} \quad \hat{f}(\mathrm{id}) = H(f)(z(f))$$

and

$$H(f)(f(\mathrm{id})) = z(\hat{f}) \quad \text{and} \quad f(\mathrm{id}) = H(\hat{f})(z(\hat{f})).$$

□

The action of $\hat{T}^{\beta}$ on the graph of H pulls back to an action σ^{β} on $\mathscr{R}$

$$\sigma^{\beta}_{g}(f) = \sigma_{H(f)^{-1}(g)}(f).$$

Defining a map $\tilde{q}$ by $\tilde{q}(f) = \hat{f}$, $\tilde{q}$ conjugates σ^{β} to $\sigma = \sigma^{\alpha}$.

On $\mathscr{R}$ of course $\boldsymbol{\beta} = \boldsymbol{\alpha\phi}$ where $\boldsymbol{\phi}(f) = \sigma_{f(\mathrm{id})}(f)$. Our work above tells us that $\boldsymbol{\alpha}(f, \boldsymbol{\phi}^{-1}(f)) = z(f)$ and more generally that

$$\boldsymbol{\alpha}(f, \boldsymbol{\phi}^{-1}(\sigma_g(f))) = z(\sigma_g(f))g = T_g(H(f))^{-1}(\hat{f}(H(\hat{f})(g)))g.$$

Notice that on the graph of H we actually have two conjugations of σ^{α} to σ^{β}. One is the map $f \to \hat{f}$, the global symmetry we have been following throughout our construction. The other is the explicit full-group element $\boldsymbol{\phi}$ pulled back from the graph of H to $\mathscr{R}$. This latter conjugation acts as the identity on orbits of σ and hence will preserve all invariant measures. The former conjugation, as we have seen, interchanges the roles of $\boldsymbol{\alpha}$ and $\boldsymbol{\beta}$. As in the case $\boldsymbol{\beta} = \boldsymbol{\alpha\phi}$ it switches the pair $\boldsymbol{\alpha}, \boldsymbol{\alpha\phi}$ to the pair $\boldsymbol{\beta}, \boldsymbol{\beta\phi}^{-1}$.

As we did with $\mathscr{G}$ we can also add a further coordinate Z, a compact metrizable space, and consider the shift acting on $\mathscr{R} \times Z^G \subseteq (G \times Z)^G$, and $Y_0 \subseteq \mathscr{R} \times Z^G$, the set of free σ-orbits in it. As the map H is 1-1 on individual orbits, $H \times \mathrm{id} : \mathscr{R} \times Z^G \to \mathscr{G} \times Z^G$ maps Y_0 to Y, the $T \times \sigma$ free orbits in $\mathscr{G} \times Z^G$ (this map is not 1-1 but is onto and continuous). The involution $\tilde{q}$ of $\mathscr{R}$ extends to this new coordinate by using the form $\hat{q}$ has on the Z^G coordinate, i.e.

$$\tilde{q}(f, \vec{z}) = (\hat{f}, \vec{z} \circ (H(f))^{-1}) = (\hat{f}, \vec{z} \circ H(\hat{f})).$$

This map $\tilde{q}$ conjugates σ^{β} to $\sigma = \sigma^{\alpha}$ on this extended system.

Given a free and ergodic G-action $(Z, \mathscr{H}, \mu, T^{\alpha}_{g})$ where Z is a compact metrizable space, and ϕ is in its full-group, to each $z \in Z$ (μ-a.s.) we can assign an $f \in \mathscr{R}$ by

$$f^z(g) = f^{\alpha,\phi}_z(g) = \alpha(T^{\alpha}_g(z), \phi(T^{\alpha}_g(z)))$$

and a Z-name $p(z)$. Let $L'(z) = (f^z, p(z)) \in Y_0 \subseteq \mathscr{R} \times Z^G$.

As expected the diagram below commutes:

$$\begin{array}{ccc} z & \xrightarrow{\quad T^{\alpha}_{g} \quad} & T^{\alpha}_{g}(z) \\ \Big\downarrow L' & & \Big\downarrow L' \\ (f^z, p(z)) & \xrightarrow{\quad \sigma^{\alpha}_{g} \quad} & \sigma^{\alpha}_{g}(f^z, p(z)) \end{array}$$

We also see that setting $\beta = \alpha\phi$, $h^z = H(f^z)$ (the bijection associated with f^z), the following diagram commutes:

$$\begin{array}{ccc} & z & \\ {}^{L}\swarrow & & \searrow^{L'} \\ (f^z, p(z)) & \xrightarrow{H\times \mathrm{id}} & (h^z, p(z)) \end{array}$$

Note that L is the map discussed preceding Lemma 6.2.3 and in this case associated to the pair of arrangements $(\alpha, \beta) = (\alpha, \alpha\phi)$.

We now begin construction of a topological model for a sequence of rearrangements converging to a second arrangement

$$\alpha\phi_i \to \beta.$$

Remember that for a measure-preserving G-action T^α and sequence of rearrangements $\{\phi_i\}$ with

$$\|\alpha\phi_i, \beta\|_1 \underset{i}{\to} 0$$

we are only obtaining $h_x^{\alpha,\alpha\phi_i} \underset{i}{\to} h_x^{\alpha,\beta}$ in L^1. Given such a sequence one can always find a subequence ψ_i with $d(h_x^{\alpha,\alpha\psi_i}, h_x^{\alpha,\beta}) \underset{i}{\to} 0$ for μ-a.e. x. It is this notion of pointwise convergence we will topologize.

What we are going to see in this construction is that it is essential to introduce invariant measures to reacquire the natural symmetry of $\alpha\phi_i \to \beta$ and $\beta\phi_i^{-1} \to \alpha$.

Definition 6.3.6. *Consider the space $\mathcal{R}^{\mathbb{N}}$ with the shift σ acting simultaneously on all copies of $\mathcal{R}$. Recall d, the metric defined in Lemma 2.1.6 making $\mathcal{G}$ a complete metric space, and for which $q : h \to h^{-1}$ acted isometrically. Let $\vec{\mathcal{R}}$ consist of those sequences $\{f_i\} \in \mathcal{R}^{\mathbb{N}}$ for which the sequence $\{H(f_i)\}$ is d-Cauchy, i.e. converges to some $h \in \mathcal{G}$. This is a σ-invariant subset of $\mathcal{R}^{\mathbb{N}}$. The map $\ell \circ H : \{f_i\} \to h$ is a σ-equivariant Borel map from $\vec{\mathcal{R}}$ to $\mathcal{G}$, mapping the sequence $\{f_i\}$ to the limit point $h = \ell(H(f_i))$. Notice that $\ell \circ H$ conjugates σ acting on $\vec{\mathcal{R}}$ to S^α acting on $\mathcal{G}$. The set $\vec{\mathcal{R}}$ is not so useful to us, most especially as the map $\{f_i\} \to h$ is not continuous.*

Let $\hat{\mathcal{R}}_I \subseteq \vec{\mathcal{R}}$ consist of those sequences $\{f_i\} \in \vec{\mathcal{R}}$ for which

$$d(H(f_i), H(f_j)) \leq 1/i + 1/j \text{ for all } i, j \geq I,$$

that is to say are uniformly Cauchy at this prescribed rate at indices beyond I.

Let $\hat{\mathcal{R}} = \cup_I \hat{\mathcal{R}}_I$.

Notice that to say

$$\limsup_{j\to\infty} d(H(f_i), H(f_j)) \leq 1/i$$

for all $i \geq I$ is equivalent to saying

$$d(H(f_i), H(f_j)) \leq 1/i + 1/j$$

for all $j \geq i \geq I$ and this in turn is equivalent to

$$d(H(f_i), \ell(H(f_k))) \leq 1/i$$

for all $i \geq I$.

Lemma 6.3.7. *Each* $\hat{\mathscr{R}}_I \subseteq \mathscr{R}^{\mathbb{N}}$ *is closed but not necessarily* σ*-invariant. The map* $\ell \circ H$ *is continuous on each* $\hat{\mathscr{R}}_I$ *and as any* $\{f_i\} \in \vec{\mathscr{R}}$ *contains a subsequence in* $\hat{\mathscr{R}}_I$*, the image under* ℓ *of each* $\hat{\mathscr{R}}_I$ *and* $\vec{\mathscr{R}}$ *are the same. (Note: this image is of course all of* $\mathscr{G}$*.)*

Proof As H is continuous, $\otimes H : \mathscr{R}^{\mathbb{N}} \to \mathscr{G}^{\mathbb{N}}$ is continuous. In $\mathscr{G}^{\mathbb{N}}$, for each value I, $\{\{h_i\} : d(h_i, h_j) \leq 1/i + 1/j, \quad j \geq i \geq I\}$ is closed, hence its inverse image, $\hat{\mathscr{R}}_I$, in $\vec{\mathscr{R}}$ is closed. As $\ell \circ H$ on $\hat{\mathscr{R}}$ is the uniform limit of the continuous maps $H_i : \{f_j\} \to H(f_i)$, it is continuous.

The possible lack of σ-invariance for $\hat{\mathscr{R}}$ is simply that S^{α} need not be a d-isometry on $\mathscr{G}$. □

For a convergent sequence of elements $h_i \to h$ in $\mathscr{G}$, clearly $h_i^{-1} \to h^{-1}$. Thus $\ell(h_i^{-1}) = \ell(h_i)^{-1}$. As q is a d-isometry, the map $\tilde{q}^{\mathbb{N}} : \{f_i\} \to \{\hat{f}_i\}$ is a homeomorphism of each $\hat{\mathscr{R}}_I$, which projects by $H^{\mathbb{N}}$ to $p^{\mathbb{N}}$ on $\mathscr{G}^{\mathbb{N}}$. It is sad, and also obvious, that $\tilde{q}^{\mathbb{N}}$ is not the conjugation of the convergent sequence $\boldsymbol{\alpha\phi} \to \boldsymbol{\beta}$ to $\boldsymbol{\beta\phi}^{-1} \to \boldsymbol{\alpha}$, that is to say, $f_{\{f_i\}}^{\boldsymbol{\beta},\boldsymbol{\phi}_i^{-1}} \neq \hat{f}_i$. The reason is that $\hat{f}_i$ describes how $\boldsymbol{\phi}_i^{-1}$ looks on the $\boldsymbol{\alpha\phi}_i$ arrangement of the orbit of f, not the $\boldsymbol{\beta}$ arrangement.

Let $\{f_i\} \in \vec{\mathscr{R}}$ and $\boldsymbol{\beta}$ be the ordering on orbits in $\vec{\mathscr{R}}$ obtained as the pull-back via $\ell \circ H$ to $\vec{\mathscr{R}}$ of the ordering $\boldsymbol{\beta}$ on $\mathscr{G}$. That is to say

$$\begin{aligned}\boldsymbol{\beta}(\{f_i\}, \sigma_g^{\alpha}(\{f_i\})) &= \lim_{j\to\infty} \boldsymbol{\alpha\phi}_j(\{f_i\}, \sigma_g(\{f_i\})) \\ &= \lim_{j\to\infty} H(f_j)(g) = \ell(H(f_j))(g) = h(g).\end{aligned}$$

Define $\{\tilde{f}_i\}$ by

$$\tilde{f}_i(g) = \boldsymbol{\beta}(\sigma_g^{\beta}(\{f_i\}), \boldsymbol{\phi}_i^{-1}(\sigma_g^{\beta}(\{f_i\}))).$$

Notice that $\tilde{f}_i$ depends on both f_i and the limit of the $H(f_i)$, and is the element of $\mathscr{R}$ describing how ϕ_i^{-1} rearranges the β-arrangement of the orbit of $\{f_i\}$.

Lemma 6.3.8. *For*

$$\beta(\{f_i\}, \sigma_g(\{f_i\})) = \ell(H(f_i))(g) \text{ and } \phi_i(\{f_i\}) = \sigma_{f_i(id)}(\{f_i\}),$$

we have

$$\tilde{f}_i(g) = \beta(\sigma_g^\beta(\{f_i\}), \phi_i^{-1}(\sigma_g^\beta(\{f_i\}))) = h(z(\sigma_{h^{-1}(g)}^\alpha(f_i))h^{-1}(g))g^{-1}$$

where, as before, $h = \ell(H(f_j))$ is the limit of the $H(f_j)$.

Proof Notice

$$\begin{aligned}\tilde{f}_i(g) &= \beta(\sigma_g^\beta(\{f_i\}), \phi_i^{-1}(\sigma_g^\beta(f_i)))\\ &= \beta(\{f_j\}, \phi_i^{-1}(\sigma_g^\beta(\{f_j\})))g^{-1}\\ &= h(\alpha(\{f_j\}, \phi_i^{-1}(\sigma_{h^{-1}(g)}^\alpha(\{f_j\}))))g^{-1}\\ &= h(z(\sigma_{h^{-1}(g)}^\alpha(f_i))h^{-1}(g))g^{-1}.\end{aligned}$$

as $\alpha(f, \phi^{-1}(\sigma_g^\alpha(f))) = z(\sigma_g^\alpha(f))g$. □

Corollary 6.3.9.

$$\begin{aligned}\beta(\phi_i^{-1}(\{f_i\}), \phi_i^{-1}(\sigma_g^\alpha(\{f_i\}))) &= \beta(\phi_i^{-1}(\{f_i\}), \phi_i^{-1}(\sigma_{h(g)}^\beta(\{f_i\})))\\ &= H(\tilde{f}_i)(h(g)).\end{aligned}$$

Proof This is a direct calculation from the definition of $\tilde{f}_i$. □

Corollary 6.3.10. *On each $\hat{\mathscr{R}}_I$, the map*

$$\overline{q} : \{f_k\} \to \{\tilde{f}_k\}$$

is continuous.

Proof Fix $g \in G$ and a value i. All we need to show is that $\tilde{f}_i(g)$ is a constant in some neighborhood of $\{f_k\}$.

Remember that

$$\tilde{f}_i(g) = h(z(\sigma_{h^{-1}(g)}^\alpha(f_i))h^{-1}(g))g^{-1},$$

where, as always, $h = \ell(H(f_k))$. We have seen that the map $\{f_k\} \to h$ is continuous on each $\hat{\mathscr{R}}_I$. We already know that the map $f \to \hat{f}$ is continuous. Hence the map $f \to z(f) = H(\hat{f})(\hat{f}(\mathrm{id}))$ is continuous, i.e. is constant in some neighborhood of each f.

The map $\{f_k\} \to h^{-1}(g)$ is continuous and so is constant in some neighborhood of each $\{f_k\}$. Hence $\{f_k\} \to \sigma^{\alpha}_{h^{-1}(g)}(f_i)$ is continuous as also will be

$$\{f_k\} \to z(\sigma^{\alpha}_{h^{-1}(g)}(f_i))h^{-1}(g).$$

Hence this choice of element in G is constant in some neighborhood of each $\{f_k\}$. Continuity of the map h on each $\hat{\mathscr{R}}_I$ now tells us that $\{f_k\} \to \tilde{f}_i(g)$ is constant on some neighborhood of each $\{f_k\}$ which is to say, is continuous. □

For actions of $\mathbb{Z}$ it is relatively easy to construct sequences $\{f_i\}$ with $H(f_i)$ convergent but with $H(\tilde{f}_i)$ failing to converge along any subsequence. In particular

$$H(\tilde{f}_i(g)) = h(z(\sigma^{\alpha}_{h^{-1}(g)}(f_i))h^{-1}(g))h(z(f_i))^{-1}$$

and

$$z(\sigma^{\alpha}_{h^{-1}(g)}(f_i)) = T_{h^{-1}(g)}(H(f_i))^{-1}(\hat{f}_i(H(f_i)(g))).$$

Hence the stability of the sequence $H(\tilde{f}_i)$ in a finite region around the identity is controlled by the stability of h along the moving sequence of values

$$H(f_i)^{-1}(\hat{f}_i(\mathrm{id})) = z(f_i).$$

This is one reason of course that the $z(f)$ were introduced, but it also indicates the path we will take in order to force the convergence of the $\boldsymbol{\beta}\boldsymbol{\phi}_i^{-1}$ back to $\boldsymbol{\alpha}$. In doing this the fact that our point spaces are Polish will be lost, and we will instead move to spaces of invariant measures supported on these point spaces which will remain Polish.

Define a sequence of sets:

$$F_I = \{\{f_k\} \in \hat{\mathscr{R}}_I : d(H(\sigma_{z(f_i)}(f_i)), H(\sigma_{z(f_i)}(f_j))) \leq 1/i + 1/j \text{ for all } j \geq i \geq I\}.$$

Notice that as $F_I \subseteq \hat{\mathscr{R}}_I$, the map $\overline{q}$ is continuous on F_I. As $\sigma_{z(f_i)} = \boldsymbol{\phi}_i^{-1}$ at the value f_i, what this is asking is that the convergence rate that is required near the identity for $\{f_k\}$ to be in $\hat{\mathscr{R}}_I$ also holds near the value $\boldsymbol{\phi}_i^{-1}(\mathrm{id})$.

Lemma 6.3.11. *The set F_I is a closed subset of $\hat{\mathscr{R}}_I$. For $\{f_k\} \in F_I$, the sequence $\{\tilde{f}_k\} \in \vec{\mathscr{R}}$, and one can consider the map*

$$\ell \circ H \circ \overline{q} : \{f_k\} \to \ell(H(\overline{q}(\{f_k\}))).$$

On each F_I this is a continuous map, and moreover on each $F_I \cap \overline{q}(F_I)$ it is a continuous involution with $\ell(H(\tilde{f}_k)) = \ell(H(f_k))^{-1}$.

Proof The map $f_i \to z(f_i)$ is continuous and hence is a constant in some neighborhood of $\{f_k\}$. That F_I is closed is now simply that it is the intersection of a collection of closed sets.

For the second part, this is most easily calculated in the vocabulary of arrangements, i.e. by writing points on the orbit of $\{f_k\} = x_0$ as x or y, setting $\boldsymbol{\alpha}(x, y) = g$ where $\sigma_g(x) = y$, $\boldsymbol{\phi}_i(y) = \sigma_{f_i(\boldsymbol{\alpha}(x_0,y))}(y)$ and $\boldsymbol{\beta}(x, y) = \lim_{i\to\infty} \boldsymbol{\alpha}(\boldsymbol{\phi}_i(x), \boldsymbol{\phi}_i(y))$. Thus $h^{\boldsymbol{\alpha},\boldsymbol{\alpha\phi}_i}_{\sigma_g(x_0)} = H(\sigma_g(f_i))$ and for $\boldsymbol{\alpha}(x_0, y) = g_0$,

$$\begin{aligned} h^{\boldsymbol{\alpha},\boldsymbol{\beta}}_y(g) &= \boldsymbol{\beta}(y, \sigma_g(y)) \\ &= \boldsymbol{\beta}(x_0, \sigma_{gg_0}(x_0))\boldsymbol{\beta}(x_0, \sigma_{g_0}(x_0)) \\ &= \lim_{i\to\infty} H(f_i)(gg_0)H(f_i)(g_0)^{-1}. \end{aligned}$$

We also know that

$$\boldsymbol{\phi}_i^{-1}(\sigma_g(x_0)) = \sigma_{z(\sigma_g(f_i))}(\sigma_g(x_0)).$$

So saying

$$d(H(\sigma_{z(f_i)}(f_i)), H(\sigma_{z(f_i)}(f_j))) \leq 1/i + 1/j$$

is to say

$$d(h^{\boldsymbol{\alpha},\boldsymbol{\alpha\phi}_i}_{\boldsymbol{\phi}_i^{-1}(x_0)}, h^{\boldsymbol{\alpha},\boldsymbol{\alpha\phi}_j}_{\boldsymbol{\phi}_i^{-1}(x_0)}) \leq 1/i + 1/j.$$

Letting $j \to \infty$ in this last expression implies:

$$d(h^{\boldsymbol{\alpha},\boldsymbol{\alpha\phi}}_{\boldsymbol{\phi}_i^{-1}(x_0)}, h^{\boldsymbol{\alpha},\boldsymbol{\beta}}_{\boldsymbol{\phi}_i^{-1}(x_0)}) \leq 1/i.$$

As d is q-invariant,

$$d(h^{\boldsymbol{\alpha\phi}_i,\boldsymbol{\alpha}}_{\boldsymbol{\phi}_i^{-1}(x_0)}, h^{\boldsymbol{\beta},\boldsymbol{\alpha}}_{\boldsymbol{\phi}_i^{-1}(x_0)}) \leq 1/i.$$

Fix a $g \in G$ and let y be such that $\boldsymbol{\beta}(x_0, y) = g$. Then $\boldsymbol{\alpha}(x_0, y) = h^{\boldsymbol{\beta},\boldsymbol{\alpha}}_{x_0}(g) = \boldsymbol{\alpha\phi}(\boldsymbol{\phi}_i^{-1}(x_0), \boldsymbol{\phi}_i^{-1}(y))$ which we will call g_1.

Thus $h^{\boldsymbol{\alpha\phi}_i,\boldsymbol{\alpha}}_{\boldsymbol{\phi}_i^{-1}(x_0)}(g_1) = \boldsymbol{\alpha}(\boldsymbol{\phi}_i^{-1}(x_0), \boldsymbol{\phi}_i^{-1}(y))$, an element of G we will call g_2.

From our statement above, once i is large enough, depending of course on g_1,

$$h^{\beta,\alpha}_{\phi_i^{-1}(x_0)}(g_1) = g_2$$

and so for i this large,

$$g_1 = h^{\alpha,\beta}_{\phi_i^{-1}(x_0)}(g_2) = \beta(\phi_i^{-1}(x_0), \phi_i^{-1}(y)) = H(\tilde{f}_i)(g).$$

Thus $H(\tilde{f}_i)(g) = h^{\beta,\alpha}_{x_0}(g) = \ell(H(f_k))^{-1}(g)$, once i is large enough. As $\{f_k\}$ and $\{\tilde{f}_k\} = \overline{q}(\{f_k\}) \in \hat{\mathscr{R}}_I$,

$$\{f_k\} \xrightarrow{\overline{q}} \{\tilde{f}_k\} \xrightarrow{\ell\circ H} \ell(H(\tilde{f}_k))$$

are continuous, finishing the result. □

The definition of the F_I, from a topological dynamical perspective, is perhaps un-natural, but as we will quickly see, very easily understood in the context of shift-invariant measures.

Consider the σ-invariant and ergodic measures $\mathscr{M}_e(\vec{\mathscr{R}})$. This is non-empty as any constructed sequence of rearrangements $(\alpha, \{\phi_i\})$ with $\alpha\phi_i$ converging in L^1 rapidly enough, in some ergodic measure-preserving G-action will project to $\vec{\mathscr{R}}$ via the map $z \to \{L'(\alpha, \phi_i)(z)\}$ and this map will carry the invariant measure for the action to $\mathscr{M}_e(\vec{\mathscr{R}})$.

Remember

$$\hat{\mathscr{R}}_I = \{\{f_i\} : d(H(f_i), H(f_j)) \leq 1/i + 1/j \text{ for all } I \geq i \geq j\},$$

a closed subset of $\vec{\mathscr{R}}$.

For any standard probability space $(X, \mathscr{F}, \mu)$ and Borel maps $x \to h_1(x), h_2(x) \in \mathscr{G}$ we can make the calculation

$$\int d(h_1(x), h_2(x))\, d\mu.$$

Notice that this is actually a calculation on $\mathscr{G}^2$ as it can be written

$$\int d(h_1, h_2)\, d(h_1 \otimes h_2)^*(\mu)$$

where

$$h_1 \otimes h_2(x) = (h_1(x), h_2(x)) \in \mathscr{G}^2.$$

On $\mathscr{G}^2$, the map $(h_1, h_2) \to d(h_1, h_2)$ is continuous, so the map

$$\mu \to \int d(h_1, h_2)\, d\mu, \text{ taking } \mathscr{M}(\mathscr{G}^2) \to [0, 1]$$

is continuous.

Definition 6.3.12. *For $\{f_j\} \in \vec{\mathscr{R}}$ define functions*

$$R_{i,J} = \max_{J \geq j \geq i} d(H(f_i), H(f_j)),$$

an increasing sequence in J, and its pointwise limit

$$R_i(\{f_j\}) = \sup_{j \geq i} d(H(f_i), H(f_j)).$$

Our above observations tell us the $R_{i,J}$ are continuous, and as d is at most one, R_i is a bounded Borel function.

Notice that if $(X, \mathscr{F}, \mu, T^\alpha)$ is a free and ergodic G-action, and $\alpha\phi_i$ converges to β in L^1 then for any sequence of values $\eta_i > 0$ we can select a subsequence ψ_i of ϕ_i with

$$\int R_i(\{f^{\alpha,\psi_i}\})\, d\mu < \eta_i.$$

Just as in the previous discussion, this is actually a calculation in $\mathscr{M}_e(\vec{\mathscr{R}})$,

$$\int R_i(\{f_j\})\, d\nu$$

where

$$\nu = L'(\alpha, \psi_i)^*(\mu) \in \mathscr{M}_e(\vec{\mathscr{R}}).$$

Definition 6.3.13. *Let*

$$\mathscr{M}_* = \{\mu \in \mathscr{M}_e(\vec{\mathscr{R}}) : \int R_i\, d\mu < (2^{i+2} i)^{-1} \textit{ for all } i\}.$$

Note the strict inequality here. It will play an essential role, and create the need for some subtlety in the following argument.

Certainly for $\mu \in \mathscr{M}_*$,

$$\mu(\{\{f_k\} : d(H(f_i), H(f_j)) \leq 1/i \text{ for all } j \geq i\}) > 1 - 2^{-(i+2)}$$

and hence

$$\mu(\hat{\mathscr{R}}_I) > 1 - 2^{-(I+1)}.$$

Thus, as ϕ_i is μ-preserving,

$$\mu(F_I) > 1 - 2^{-I}.$$

This is where the role of invariant measures comes into play. If $\mu(\hat{\mathscr{R}}_I)$ is close to one then $\mu(F_I)$ is also automatically close to one. Hence for

μ-a.e. $\{f_k\}$, $\{f_k\} \in F_I$ for all sufficiently large I, and so for $\mu \in \mathcal{M}_*$, for all i,

$$\int d(H(f_i), \ell(H(f_k)))\, d\mu \leq \int R_i\, d\mu \leq (2^{i+2}i)^{-1}.$$

Thus by restricting to measures in $\mathcal{M}_*$ our measures will all be supported on the lim sup of the F_I's. This is an F_σ and hence not a Polish space, but as our interest has shifted to spaces of measures this will not matter. We call this set $\hat{F}$, i.e.

$$\cup_I \cap_{i \geq I} F_i = \hat{F}.$$

Lemma 6.3.14. *$\mathcal{M}_*$ is a G_δ subset of $\mathcal{M}_e(\vec{\mathcal{R}})$.*

Proof For each $i \leq J$,

$$\{\mu : \int R_{i,J}\, d\mu \leq (2^{i+2}i)^{-1}\}$$

is a closed set of measures and as $R_{i,J} \nearrow_J R_i$ pointwise,

$$\mathcal{M}' = \{\mu : \int R_i\, d\mu \leq (2^{i+2}i)^{-1} \text{ for all } i\}$$

is a closed set of measures containing $\mathcal{M}_*$. Restrict the topology and the discussion to this subset.

For $\mu \in \mathcal{M}_*$ and i fixed, there is a value $\varepsilon_i(\mu) > 0$ with

$$\int R_i\, d\mu < (2^{i+2}i)^{-1} - \varepsilon_i(\mu).$$

Choose $J = J(i, \mu)$ so large that for all $\mu' \in \mathcal{M}'$,

$$\int R_J\, d\mu' \leq (2^{J+2}J)^{-1} < \varepsilon(\mu)/2.$$

As $R_{i,J(i,\mu)}$ is continuous, for some neighborhood $\mathcal{O}(i,\mu)$ of μ, for $\mu' \in \mathcal{O}(i,\mu)$, we will still have

$$\int R_{i,J(i,\mu)}\, d\mu' < (2^{i+2}i)^{-1} - \varepsilon_i(\mu).$$

Hence (using the fact that $\mu' \in \mathcal{M}'$)

$$\int R_i\, d\mu' \leq \int (R_{i,J(i,\mu)} + R_{J(i,\mu)})\, d\mu' < (2^{i+2}i)^{-1} - \varepsilon_i(\mu)/2.$$

Let $\mathcal{O}_i = \cup_{\mu \in \mathcal{M}_*} \mathcal{O}(i,\mu)$, an open set. For any $\mu' \in \mathcal{O}_i$ we have

$$\int R_i\, d\mu < (2^{i+2}i)^{-1}$$

(a strict inequality) and so

$$\mathcal{M}_* \subseteq \cup_{\mu \in \mathcal{M}_*} \mathcal{O}_i \subseteq \mathcal{M}_*.$$

□

For any $\mu \in \mathcal{M}_*$ we know that $\mu(F_I) \geq 1 - 2^I$. On each F_I the map $\overline{q}$ is continuous. These two facts imply that $\overline{q}^*$ acts continously on the space of measures $\mathcal{M}_*$

$$\overline{q}^* : \mathcal{M}_* \to \mathcal{M}_e(\vec{\mathcal{R}}).$$

Definition 6.3.15. *Set*

$$\mathcal{M}_{**} = \{\mu \in \mathcal{M}_* : \overline{q}^*(\mu) \in \mathcal{M}_*\}.$$

Since for $\mu \in \mathcal{M}_{**}$ we have $\mu(F_I \cap \overline{q}(F_I)) \geq 2^{-I+1}$, and on each $F_I \cap \overline{q}(F_I)$, $\overline{q}$ acts as a continuous involution, $\overline{q}^*$ acts as a continous involution on $\mathcal{M}_{**}$.

Lemma 6.3.16. *$\mathcal{M}_{**}$ is a G_δ subset of $\mathcal{M}_*$ and hence is a Polish space of measures.*

Proof As $\overline{q}^*$ acts continuously on $\mathcal{M}_*$,

$$\mathcal{M}' = \{\mu \in \mathcal{M}_* : \int R_i \, d\overline{q}^*(\mu) \leq (2^{i+2}i)^{-1} \text{ for all } i\}$$

is a closed subset of $\mathcal{M}_*$ containing $\mathcal{M}_{**}$. We restrict the discussion and the topology to this closed subset. For $\mu' \in \mathcal{M}'$,

$$\mu(\overline{q}(F_I)) \geq 1 - 2^{-I}$$

for all I. For $\mu \in \mathcal{M}_{**}$ and i fixed, there is a value $\varepsilon_i(\mu) > 0$ with

$$\int R_i \, d\overline{q}^*(\mu) < (2^{i+2}i)^{-1} - \varepsilon_i(\mu).$$

Choose $I = I(i, \mu)$ sufficiently large so that for all $\mu' \in \mathcal{M}'$,

$$\mu'(\overline{q}(F_I)) > 1 - 2^{-I} > 1 - \varepsilon_i(\mu)/3$$

and further,

$$\int R_I \, d\overline{q}^*(\mu') \leq (2^{I+2}I)^{-1} < \varepsilon_i(\mu)/3.$$

Choose a neighborhood $\mathcal{O}(i,\mu)$ so that for $\mu' \in \mathcal{O}(i,\mu)$ we still have

$$\int_{\overline{q}(F_{I(i,\mu)})} R_{i,J(i,\mu)}\, d\overline{q}^*(\mu') = \int_{F_{I(i,\mu)}} (R_{i,J(i,\mu)} \circ \overline{q})\, d\mu' < (2^{i+2}i)^{-1} - \varepsilon_i(\mu).$$

As all the $R_{i,J}$ are bounded by 1 we conclude

$$\begin{aligned}\int R_i\, d\overline{q}^*(\mu') &\leq \int_{F_{I(i,\mu)}} (R_{i,J(i,\mu)} + R_{I(i,\mu)})\, d\overline{q}^*(\mu') + \mu'(\overline{q}(F^c_{I(i,\mu)}))\\ &\leq (2^{i+2}i)^{-1} - \varepsilon_i(\mu)/3.\end{aligned}$$

Setting $\mathcal{O}_i = \cup_{\mu\in\mathcal{M}_*}\mathcal{O}(i,\mu)$, an open set, for $\mu' \in \mathcal{O}_i$,

$$\int R_i\, d\overline{q}^*(\mu') < (2^{i+1}i)^{-1},$$

a strict inequality. Hence

$$\mathcal{M}_{**} \subseteq \cap_i \mathcal{O}_i \subseteq \mathcal{M}_{**}.$$

□

To help set the current situation in mind we state the following corollary.

Corollary 6.3.17. *The map $\overline{q}$ is of order two on each $\{f_k\} \in \hat{F}$, (a set given full measure by any measure in $\mathcal{M}_{**}$) and $\overline{q}^*$ is a continuous involution on $\mathcal{M}_{**}$. Moreover the diagrams below commute:*

$$\begin{array}{ccc} \hat{F} & \xrightarrow{\ \overline{q}\ } & \hat{F} \\ \ell\circ H \downarrow & & \downarrow \ell\circ H \\ \mathcal{G} & \xrightarrow{\ q\ } & \mathcal{G} \end{array} \qquad \begin{array}{ccc} \mathcal{M}_{**} & \xrightarrow{\ \overline{q}^*\ } & \mathcal{M}_{**} \\ \ell\circ H^* \downarrow & & \downarrow \ell\circ H^* \\ \mathcal{M}_e(\mathcal{G}) & \xrightarrow{\ q^*\ } & \mathcal{M}_e(\mathcal{G}) \end{array}$$

$\mathcal{M}_{**}$ is not empty since for any G-action $(Z, \mathcal{H}, \mu, T^\alpha)$ and sequence of full-group elements ϕ_i, with

$$\alpha\phi_i \underset{i}{\rightarrow} \beta,$$

in L^1 there is a subsequence $\psi_i = \phi_{t(i)}$ so that defining an image of the subsequence in $\mathcal{R}^{\mathbb{N}}$ by

$$\hat{L}_{\alpha,\psi_i}(z) = \{f_z^{\alpha,\psi_i}\},$$

then $\hat{L}^*_{\alpha,\psi_i}(\mu) \in \mathcal{M}_{**}$.

Proof All but the last statement have been proven earlier. For the last part of the corollary, just notice that if

$$\|\alpha\psi_i, \beta\|_1$$

is sufficiently small, then

$$d(h_x^{\alpha,\alpha\psi_i}, h_x^{\alpha,\beta}) < (2^{i+4}i)^{-1}$$

for a set of points x of μ-measure strictly greater than $1 - (2^{i+4}i)^{-1}$, and similarly for those x with

$$d(h_x^{\beta,\beta\psi_i^{-1}}, h_x^{\beta,\alpha}) < (2^{i+4}i)^{-1}.$$

It is now a relatively simple calculation to show that $\hat{L}^*_{\alpha,\psi_i}(\mu) \in \mathcal{M}_{**}$. □

As we did before with $\mathcal{G}$ and $\mathcal{R}$, we can again extend $\vec{\mathcal{R}}$ to $\vec{\mathcal{R}} \times Z^G$ and consider the σ-invariant, ergodic and aperiodic measures which project to an element of $\mathcal{M}_{**}$. We call these measures $\mathcal{M}_{**}(Z)$. The action of $\ell \circ H$ and $\bar{q}$ extend to this expanded space, as before, by rearranging the Z^G-name precisely as the G^G-name is rearranged. Any measure $\mu \in \mathcal{M}_{**}(Z)$ will project, by the trivial extension of $\ell \circ H$ to Z-names, to $\mathcal{M}_e(\mathcal{G} \times Z^G)$. Furthermore the continuous involution $\bar{q}$ will extend naturally to the Z-names and the following diagram commutes.

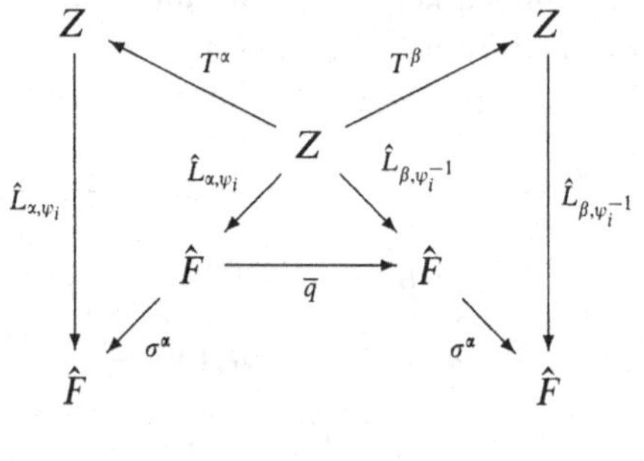

Corollary 6.3.18. *The space $\mathcal{M}_{**}(Z)$ is a Polish space, $\bar{q}$ acts on it as a continuous involution and $\ell \circ H^*$ projects it continuously into $\mathcal{M}_e(Z)$.*

Proof $\vec{\mathcal{R}}$ is a Polish space, and so the free σ-orbits on $\vec{\mathcal{R}} \times Z^G$ are a Polish space by Lemma 6.1.9. Hence the σ-invariant and ergodic measures on this space are a Polish space. Those measures whose restriction $\vec{\mathcal{R}}$ lie in $\mathcal{M}_{**}$ are, as we have seen, a closed set, and hence form a Polish space.

The rest of the corollary follows as a direct consequence of the parallel results on $\mathcal{M}_{**}$. □

6.4 Adding sizes to the picture

We now bring sizes into the picture developed in the previous Section. On the space $Z^G \times \vec{\mathscr{R}}$ we have two canonical arrangements $\boldsymbol{\alpha}$ and $\boldsymbol{\beta}$ and a canonical sequence of elements $\boldsymbol{\phi}_i$ with

$$\boldsymbol{\alpha}\boldsymbol{\phi}_i \underset{i}{\to} \boldsymbol{\beta}.$$

In order for a size m to provide an evaluation on a rearrangement $(\boldsymbol{\alpha}\boldsymbol{\phi}_i, \boldsymbol{\phi}_i^{-1}\boldsymbol{\phi}_j)$ we must specify a measure making the action of $\sigma^{\boldsymbol{\alpha}}$ free and ergodic. On the other hand, if we are given such a measure we can immediately make the evaluation of the size m of the rearrangement. We include this dependence of the evaluation on μ by writing

$$m_\mu(\boldsymbol{\alpha}\boldsymbol{\phi}_i, \boldsymbol{\phi}_i^{-1}\boldsymbol{\phi}_j).$$

Definition 6.4.1. *Let m be a G-size and Z a Polish space. Define*

$$\mathscr{M}_0^m(Z) = \{\mu \in \mathscr{M}_{**}(Z) : \limsup_{j\to\infty} m_\mu(\boldsymbol{\alpha}\boldsymbol{\phi}_i, \boldsymbol{\phi}_i^{-1}\boldsymbol{\phi}_j) < 1/i \text{ for all } i\}.$$

Let

$$\mathscr{M}^m(Z) = \{\mu \in \mathscr{M}_{**}(Z) : \text{both } \mu \text{ and } \overline{q}(\mu) \text{ are in } \mathscr{M}_0^m(Z)\}.$$

Notice that for any G-action $(Z, \mathscr{F}, \mu, T^\alpha)$ and any sequence of rearrangements $\alpha\phi_i \to \beta$ pointwise where the ϕ_i are m_α-Cauchy, there will be a subsequence of rearrangements $\phi_{t(i)}$ whose image under $\hat{L}^*_{\alpha,\phi_{t(i)}}$ will lie in $\mathscr{M}^m(Z)$. What the definition requires is a rate of convergence in m for the sequence, completely analogous to the rate required in order for a sequence $\{f_i\}$ to be in $\hat{\mathscr{R}}$. Notice that there we use non-strict inequalities, but here they are strict. This small change will play a serious role later on.

Theorem 6.4.2. *$\mathscr{M}^m(Z)$ is a G_δ subset of $\mathscr{M}_{**}(Z)$ and hence is a Polish space of measures.*

Proof For convenience let D be a metric on $\mathscr{M}_{**}(Z)$ making it a complete metric space. For each $\mu \in \mathscr{M}_0^m(Z)$ and i there is an $\varepsilon_i(\mu) > 0$ with

$$\limsup_{j\to\infty} m_\mu(\boldsymbol{\alpha}_0\boldsymbol{\phi}_i, \boldsymbol{\phi}_i^{-1}\boldsymbol{\phi}_j) < 1/i - \varepsilon_i(\mu).$$

Set $n_i(\mu) = [8/\varepsilon_i(\mu)] + 1$.

For $\mu \in \mathscr{M}_0^m(Z)$, for all i, j and k,

$$m_\mu(\boldsymbol{\alpha}\boldsymbol{\phi}_i, \boldsymbol{\phi}_i^{-1}\boldsymbol{\phi}_j) < m_\mu(\boldsymbol{\alpha}\boldsymbol{\phi}_i, \boldsymbol{\phi}_i^{-1}\boldsymbol{\phi}_k) + m_\mu(\boldsymbol{\alpha}\boldsymbol{\phi}_j, \boldsymbol{\phi}_j^{-1}\boldsymbol{\phi}_k)$$

and taking a lim sup in k,

$$m_\mu(\alpha\phi_i, \phi_i^{-1}\phi_j) \le 1/i + 1/j.$$

This implies that in fact $\limsup_{j\to\infty} m_\mu(\alpha\phi_i, \phi_i^{-1}\phi_j)$ must be a limit.

More importantly to us, for all $j > n_i(\mu)$ we must have

$$m_\mu(\alpha\phi_i, \phi_i^{-1}\phi_j) < 1/i - 3\varepsilon_i(\mu)/4.$$

(Otherwise the lim sup would be too large.)

Choose radii $r_I(\mu) < 2^{-I}$ so that if $D(\mu, \mu') < r_I(\mu)$ then by Axiom 3, for all $i = 1, 2, \ldots, I$ and each $j = n_i(\mu)$

$$m_{\mu'}(\alpha\phi_i, \phi_i^{-1}\phi_j) < 1/i - 3\varepsilon_i(\mu)/4.$$

Define open sets

$$\mathcal{O}_I = \bigcup_{\mu \in \mathcal{M}_0^m(Z)} B_{r_I(\mu)}(\mu).$$

Obviously

$$\mathcal{M}_0^m(Z) \subseteq \cap_I \mathcal{O}_I.$$

To see the other containment, suppose $\mu \in \cap_I \mathcal{O}_I$.

For each I then there must exist $\mu_I \in \mathcal{M}_0^m(Z)$ with

$$D(\mu, \mu_I) < r_I(\mu_I) < 2^{-I}.$$

We can then find a subsequence $I(i) \ge i$ so that

$$B_{r_{I(i+1)}(\mu_{I(i+1)})}(\mu_{I(i+1)}) \subseteq B_{r_{I(i)}(\mu_{I(i)})}(\mu_{I(i)}).$$

From this, for all $k \ge i$, and $j \ge n_i(\mu_{I(i)})$, setting $j' = n_i(\mu_{I(i)})$,

$$m_{\mu_{I(k)}}(\alpha\phi_i, \phi_i^{-1}\phi_j) \le m_{\mu_{I(k)}}(\alpha\phi_i, \phi_i^{-1}\phi_{j'}) + m_{\mu_{I(k)}}(\alpha\phi_{j'}, \phi_{j'}^{-1}\phi_j).$$

As $\mu_{I(k)} \in B_{r_{I(i)}(\mu_{I(i)})}(\mu_{I(i)})$ and also is in $\mathcal{M}_0^m(Z)$, this value is

$$\begin{aligned} < 1/i - 3\varepsilon_i(\mu_{I(i)})/4 + 1/j + 1/j' &\le 1/i - 3\varepsilon_i(\mu_{I(i)})/4 + \varepsilon_i(\mu_{I(i)})/4 \\ &= 1/i - \varepsilon_i(\mu_{I(i)})/2. \end{aligned}$$

Fixing i and j in this, letting $k \to \infty$ and again applying Axiom 3, we conclude

$$m_\mu(\alpha\phi_i, \phi_i^{-1}\phi_j) \le 1/i - \varepsilon_i(\mu_{I(i)})/2.$$

Examining the lim sup as $j \to \infty$ we are forced to conclude that $\mu \in \mathcal{M}_0^m(Z)$. □

Definition 6.4.3. *We say $\mu \in \mathcal{M}_{**}$ belongs to $\mathcal{M}^m$ if for some Polish space Z and $\mu_1 \in \mathcal{M}^m(Z)$, the projection of μ_1 onto its $\vec{\mathcal{R}}$ coordinates is μ.*

Corollary 6.4.4. *A measure $\mu_1 \in \mathcal{M}_e(Z^G \times \vec{\mathcal{R}})$ is in $\mathcal{M}^m(Z)$ if and only if its projection on $\vec{\mathcal{R}}$ is in $\mathcal{M}^m$. Moreover, $\mathcal{M}^m$ is a G_δ subset of $\mathcal{M}_{**}$ and hence is a Polish space.*

Proof The first part of the corollary is simply the observation that Axiom 3 tells us that $m_{\mu_1}(\alpha\phi_i, \phi_i^{-1}\phi_j)$ depends only on the projection μ of μ_1 on $\vec{\mathcal{R}}$. The proof of Theorem 6.4.2 defined its open sets $\mathcal{O}_I$ using only Axiom 3 and these values, hence it applies equally well to $\mathcal{M}^m$ as a subset of $\mathcal{M}_{**}$. □

Lemma 6.4.5. *For $\mu \in \mathcal{M}_0^m(Z)$ we know $\boldsymbol{\alpha\phi}_i \underset{m}{\rightarrow} \boldsymbol{\beta}$ relative to the measure μ as $h_z^{\boldsymbol{\alpha},\boldsymbol{\alpha\phi}_i} \rightarrow h_z^{\boldsymbol{\alpha},\boldsymbol{\beta}}$ on the support of μ and $\{\boldsymbol{\phi}_i\}$ is $m_{\mu,\boldsymbol{\alpha}}$-Cauchy. Such a μ is in $\mathcal{M}^m(Z)$ if and only if $\{\boldsymbol{\phi}_i^{-1}\}$ is $m_{\mu,\boldsymbol{\beta}}$-Cauchy, which is true if and only if $\{\boldsymbol{\phi}_i\}$ is $m_{\overline{q}^*(\mu),\boldsymbol{\alpha}}$-Cauchy.*

Proof Only one piece of this is less than direct and it requires a little calculation. In particular, we must show that if $\mu \in \mathcal{M}_0^m(Z)$ and $\{\boldsymbol{\phi}_i^{-1}\}$ is $m_{\mu,\boldsymbol{\beta}}$-Cauchy, then $\overline{q}^*(\mu)$ is in $\mathcal{M}_0^m(Z)$, that is to say $\{\boldsymbol{\phi}_i^{-1}\}$ is $m_{\mu,\boldsymbol{\beta}}$-Cauchy at a fast enough rate. To see this, let $\langle\boldsymbol{\phi}_i\rangle = \hat{\boldsymbol{\beta}}$ in the $m_{\mu,\boldsymbol{\alpha}}$-completion of the full-group. That $\mu \in \mathcal{M}_0^m(Z)$ says simply that for all i,

$$m_{\mu,\boldsymbol{\alpha}}(\boldsymbol{\phi}_i, \hat{\boldsymbol{\beta}}) < 1/i.$$

If $\{\boldsymbol{\phi}_i^{-1}\}$ is also known to be $m_{\mu,\boldsymbol{\beta}}$-Cauchy then by Theorem 2.2.7, setting $\langle\boldsymbol{\phi}_i^{-1}\rangle = \hat{\boldsymbol{\alpha}}$ in the $m_{\mu,\boldsymbol{\beta}}$-completion of the full-group, we will have

$$\begin{aligned} m_{\mu,\boldsymbol{\alpha}}(\boldsymbol{\phi}_i, \hat{\boldsymbol{\beta}}) &= m_{\mu,\boldsymbol{\beta}}(\hat{\boldsymbol{\alpha}}\boldsymbol{\phi}_i, \mathrm{id}) = m_{\mu,\boldsymbol{\beta}}(\hat{\boldsymbol{\alpha}}, \boldsymbol{\phi}_i^{-1}) \\ &= m_{\mu,\boldsymbol{\beta}}(\boldsymbol{\phi}_i^{-1}, \hat{\boldsymbol{\alpha}}) = m_{\overline{q}^*(\mu),\boldsymbol{\alpha}}(\boldsymbol{\phi}_i, \hat{\boldsymbol{\beta}}). \end{aligned}$$

Hence for all i,

$$\limsup_{j\to\infty} m_\mu(\boldsymbol{\alpha\phi}_i, \boldsymbol{\phi}_i^{-1}\boldsymbol{\phi}_j) = \limsup_{j\to\infty} m_{\overline{q}^*(\mu)}(\boldsymbol{\alpha\phi}_i, \boldsymbol{\phi}_i^{-1}\boldsymbol{\phi}_j)$$

forcing $\overline{q}^*(\mu)$ into $\mathcal{M}_0^m(Z)$. □

Definition 6.4.6. *To $\mu \in \mathcal{M}_0^m(Z)$ we can assign a value we will call $m(\mu)$ defined by*

$$m(\mu) = m_{\mu,\boldsymbol{\alpha}}(\mathrm{id}, \overline{\boldsymbol{\beta}}) = \lim_{i\to\infty} m_\mu(\boldsymbol{\alpha}, \boldsymbol{\phi}_i).$$

If in fact $\mu \in \mathcal{M}^m(Z)$ then we will also have

$$m(\mu) = m(\overline{q}^*(\mu)).$$

Corollary 6.4.7. *The function m acting on $\mathcal{M}_0^m(Z)$ is upper semi-continuous.*

Proof Axiom 3 tells us that for each ϕ_i, $m_\mu(\alpha, \phi_i)$ is upper semi-continuous in μ. As $\mu \in \mathcal{M}_0^m(Z)$, $m_\mu(\alpha\phi_i, \phi_i^{-1}\phi_j) \leq 1/i + 1/j$ for all $j \geq i$. Hence if $\mu_k \to \mu$ all in $\mathcal{M}_0^m(Z)$, then

$$\begin{aligned} \limsup_{k\to\infty}(m(\mu_k)) &= \limsup_{k\to\infty}(\lim_{i\to\infty} m_{\mu_k}(\alpha, \phi_i)) \\ &\leq \lim_{i\to\infty}(\limsup_{k\to\infty}(m_{\mu_k}(\alpha, \phi_i) + 2/i)) \leq \lim_{i\to\infty} m_\mu(\alpha, \phi_i) = m(\mu). \end{aligned}$$

□

6.5 More orbit joinings and m-joinings

Earlier we defined an orbit joining of

$$(Z_1, \mathcal{F}_1, \mu_1, T_1^{\alpha_1}) \text{ and } (Z_2, \mathcal{F}_2, \mu_2, T_2^{\alpha_2})$$

as an $S^\alpha \times \sigma \times \sigma$ invariant and ergodic probability measure $\hat{\mu}$ on $\mathcal{G} \times Z_1^G \times Z_2^G$ whose projection on Z_1^G was μ_1 mapped by p_1^* to the space of Z_1-names, and for which $q^*(\mu)$ was μ_2 mapped by p_2^* to the space of Z_2-names. We can now refine this idea to include a fixed sequence of rearrangements leading from the one action to the other in the joining.

Definition 6.5.1. *Suppose $(Z_1, \mathcal{F}_1, \mu_1, T_1^{\alpha_1})$ and $(Z_2, \mathcal{F}_2, \mu_2, T_2^{\alpha_2})$ are two free ergodic G-actions on Polish spaces. A* **C-orbit joining** *is a measure $\hat{\mu} \in \mathcal{M}_{**}(Z_1 \times Z_2)$ with*

$$\pi_1^*(\hat{\mu}) = p_1^*(\mu_1) \qquad \text{and} \qquad \pi_2^*(\overline{q}(\hat{\mu})) = p_2^*(\mu_2).$$

We write this set of measures as $J_C(T_1^{\alpha_1}, T_2^{\alpha_2})$. This is a closed subset of $\mathcal{M}_{**}$ as $\tilde{L}^*$ acts continuously on $\mathcal{M}_{**}$. It is non-empty as it contains all ergodic joinings of the two systems, extended by a point mass supported on the sequence $\{\text{id}\}$ in $\vec{\mathcal{R}}$.

The C in this definition can be taken to stand for Cauchy, or constructive, or convergent, or completion as you will, as what is modeled here is not only a pair of orderings α and β with projections to the two original actions, but an explicit sequence of full-group elements ϕ_i with $\alpha\phi_i \to \beta$.

We now want to see that $J_C(T_1^{\alpha_1}, T_2^{\alpha_2})$ as a topological space, is a conjugacy invariant of the pair of actions, that is to say, does not depend on the topology on Z_1 and Z_2 but merely on the measure-algebra structure on them. This takes us on a small digression.

For $(Z, \mathscr{F}, \mu)$ a regular probability space, let $D(Z)$ be the set of all real-valued Borel functions with $\|g\|_\infty \leq 1$. Topologize $D(Z)$ with the $L^1(\mu)$-metric making $D(Z)$ a complete separable metric space. (As usual one identifies two functions here if they are a.s. equal.)

Let $(Z_1, \mathscr{F}_1, \mu_1)$ and $(Z_2, \mathscr{F}_2, \mu_2)$ be two standard probability spaces and $\mathscr{C}(\mu_1, \mu_2)$ consists of all Borel measures on $Z_1 \times Z_2$ which project to μ_1 and μ_2 respectively on the two coordinates, topologized with the weak*-topology. We refer to measures in $\mathscr{C}(\mu_1, \mu_2)$ as couplings of μ_1 and μ_2.

On $D(Z_1) \times D(Z_2)$ place the metric

$$\|(g_1, g_2), (g_1', g_2')\| = \|g_1 - g_1'\|_1 + \|g_2 - g_2'\|_1.$$

For $\hat{\mu} \in \mathscr{C}(\mu_1, \mu_2)$ define $F_{\hat{\mu}} : D(Z_1) \times D(Z_2) \to \mathbb{R}$ by

$$F_{\hat{\mu}}(g_1, g_2) = \int g_1 \otimes g_2 \, d\hat{\mu}.$$

On $\{F_{\hat{\mu}}\}$ place the Tychanoff topology of pointwise convergence.

Lemma 6.5.2.

(i) *The functions $\{F_{\hat{\mu}}\}$ are a uniformly equi-continuous family and are uniformly bounded, hence precompact. The Tychanoff topology is metrizable as $D(Z_1) \times D(Z_2)$ is separable.*

(ii) *The map $\hat{\mu} \to F_{\hat{\mu}}$ is a homeomorphism from $\mathscr{C}(\mu_1, \mu_2)$ to $\{F_{\hat{\mu}}\}$ and hence the set $\{F_{\hat{\mu}}\}$ is compact.*

Proof That the $F_{\hat{\mu}}$ are uniformly equi-continuous is simply that

$$\begin{aligned}
|F_{\hat{\mu}}(g_1, g_2) - F_{\hat{\mu}}(g_1', g_2')| &= |\int g_1 \otimes g_2 \, d\hat{\mu} - \int g_1' \otimes g_2' \, d\hat{\mu}| \\
&\leq |\int (g_1 - g_1') \otimes g_2 \, d\hat{\mu}| + |\int g_1' \otimes (g_2 - g_2') \, d\hat{\mu}| \\
&\leq \|g_1 - g_1'\|_1 + \|g_2 - g_2'\|_1,
\end{aligned}$$

i.e. all the $F_{\hat{\mu}}$'s are "contractions". That $\|F_{\hat{\mu}}\|_\infty = 1$ completes the proof of (i).

To see (ii) we show that $\hat{\mu}_i \to \hat{\mu}$ in $\mathscr{C}(\mu_1, \mu_2)$ iff $F_{\hat{\mu}_i} \to F_{\hat{\mu}}$ pointwise.

Both directions of this are easy. If $F_{\hat{\mu}_i} \to F_{\hat{\mu}}$ then for all continuous functions $g_1 \otimes g_2$,

$$\int g_1 \otimes g_2 \, d\hat{\mu}_i \to \int g_1 \otimes g_2 \, d\hat{\mu}.$$

But then $\int g \, d\hat{\mu}_i \to \int g \, d\hat{\mu}$ for any g in the closure in the uniform topology of the linear span of all such continuous $g_1 \otimes g_2$. This uniform closure is precisely all continuous functions on $Z_1 \times Z_2$ and so $\hat{\mu}_i \to \hat{\mu}$ weak*.

Now suppose $\hat{\mu}_i \to \hat{\mu}$ weak*. Then $\int g_1 \otimes g_2 \, d\hat{\mu}_i \to \int g_1 \otimes g_2 \, d\hat{\mu}$ for all pairs of continuous functions $(g_1, g_2) \in D(Z_1) \times D(Z_2)$. The continuous functions are dense in each $D(Z_i)$, and so uniform equi-continuity of the family $\{F_{\hat{\mu}}\}$ tells us that $F_{\hat{\mu}_i} \to F_{\hat{\mu}}$. □

Corollary 6.5.3. *Suppose $(Z_i, \mathcal{F}_i, \mu_i)$ and $(Z_i', \mathcal{F}_i', \mu_i')$, $i = 1, 2$, are regular spaces, and Φ_i are two measure-preserving invertible Borel maps defined μ_i-a.e. to μ_i'-a.a. of Z_i'. Then the map $\hat{\mu} \to (\Phi_1 \times \Phi_2)^*(\hat{\mu})$ is a homeomorphism from $\mathcal{C}(\mu_1, \mu_2)$ to $\mathcal{C}(\mu_1', \mu_2')$.*

Proof The map $(g_1, g_2) \to (g_1 \circ \Phi_1, g_2 \circ \Phi_2)$ is an isometry taking

$$D(Z_1) \times D(Z_2) \to D(Z_1') \times D(Z_2').$$

Hence it takes $\{F_{\hat{\mu}'} : \hat{\mu}' \in \mathcal{C}(\mu_1' \mu_2')\}$ to some uniformly equi-continuous family of functions on $D(Z_1) \times D(Z_2)$. We want to show that this family is exactly $\{F_{\hat{\mu}} : \hat{\mu} \in \mathcal{C}(\mu_1, \mu_2)\}$. Where they are defined these maps are bounded and linear. Hence on pairs of continuous functions (g_1, g_2), each image function $F_{\hat{\mu}'} \circ (\Phi_1 \times \Phi_2)$ is the restriction of a measure. As the family of functions is uniformly equi-continuous, and the continuous functions are dense in $D(Z_1) \times D(Z_2)$, each $F_{\hat{\mu}'} \circ (\Phi_1 \times \Phi_2)$ is in fact an integral with respect to a measure. That is to say, the map $F_{\hat{\mu}'} \to F_{\hat{\mu}'} \circ (\Phi_1 \times \Phi_2)$ takes $\{F_{\hat{\mu}'} : \hat{\mu}' \in \mathcal{C}(\mu_1', \mu_2')\}$ to $\{F_{\hat{\mu}} : \hat{\mu} \in \mathcal{C}(\mu_1, \mu_2)\}$. Invertibility of the Φ_i's makes this map invertible. It is very easy to see that this map is continuous in the Tychanoff topology. Lemma 6.5.2 now completes the result. □

This corollary already tells us that the space of joinings of two actions, as a topological space, is independent of the choice of regular Borel model we choose for the two actions, as the space of joinings is a closed subset of the space of couplings of the two measures. To show this same fact for orbit-joinings, C-orbit joinings, and later *m*-joinings (*m* a size), we must discuss the extra baggage these spaces carry in the form of the Polish spaces $\mathcal{G}$ and $\vec{\mathcal{R}}$. This explains the need for our next digression.

For regular spaces $(Z_i, \mathscr{F}_i, \mu_i)$, $i = 1, 2$, and Z_3 a Polish topological space, consider

$$D(Z_1) \times D(Z_2) \times C_1(Z_3)$$

where $C_1(Z_3)$ is the space of all continuous real-valued functions of sup-norm at most 1 (the unit ball in the continuous functions). On this space put the metric

$$\|(g_1, g_2, g_3), (g_1', g_2', g_3')\| = \|g_1 - g_1'\|_1 + \|g_2 - g_2'\|_1 + \|g_3 - g_3'\|_\infty.$$

This makes $D(Z_1) \times D(Z_2) \times C_1(Z_3)$ a complete metric space.

Let $\mathscr{C}(\mu_1, \mu_2, Z_3)$ consist of all Borel measures on $Z_1 \times Z_2 \times Z_3$ whose projections on Z_1 and Z_2 are μ_1 and μ_2 respectively. For $\hat{\mu} \in \mathscr{C}(\mu_1, \mu_2, Z_3)$ let

$$F_{\hat{\mu}} : D(Z_1) \times D(Z_2) \times C_1(Z_3) \to [-1, 1]$$

be given by

$$F_{\hat{\mu}}(g_1, g_2, g_3) = \int g_1 \otimes g_2 \otimes g_3 \, d\hat{\mu}.$$

Topologize the space of all such $F_{\hat{\mu}}$'s with the Tychanoff topology of pointwise convergence.

Lemma 6.5.4.

(i) *The family of functions* $\{F_{\hat{\mu}} : \hat{\mu} \in \mathscr{C}(\mu_1, \mu_2, Z_3)\}$ *is uniformly equi-continuous and metrizable hence precompact.*

(ii) *The map* $\hat{\mu} \to F_{\hat{\mu}}$ *is a homeomorphism between* $\mathscr{C}(\mu_1, \mu_2, Z_3)$ *and* $\{F_{\hat{\mu}}\}$. *Hence* $\{F_{\hat{\mu}}\}$ *in the Tychanoff topology is a Polish space.*

Proof That the family $\{F_{\hat{\mu}}\}$ is uniformly equi-continuous is virtually identical to the same argument in Lemma 6.5.2, as is showing that $\hat{\mu}_i \to \hat{\mu}$ iff $F_{\hat{\mu}_i} \to F_{\hat{\mu}}$ pointwise. The lemma is finished by observing that $\mathscr{C}(\mu_1, \mu_2, Z_3)$ is a closed subset of the Polish space of all Borel probability measures on $Z_1 \times Z_2 \times Z_3$ and hence is a Polish space. □

Corollary 6.5.5. *Suppose* $(Z_1, \mathscr{F}_1, \mu_1, T_1^{\alpha_1})$ *and* $(Z_2, \mathscr{F}_2, \mu_2, T_2^{\alpha_2})$ *are two free measure-preserving and ergodic G-actions on regular probability spaces. Remember the space of orbit-joinings of* $T_1^{\alpha_1}$ *and* $T_2^{\alpha_2}$ *is the space of* $\hat{T}^{\alpha}$ *invariant measures* $\hat{\mu}$ *on* $\mathscr{G} \times Z_1^G \times Z_2^G$ *with* $\pi_1^*(\hat{\mu}) = p_1^*(\mu_1)$ *and* $\pi_2^*(\overline{q}^*(\hat{\mu})) = p_2^*(\mu_2)$.

This Polish space of measures in the weak-topology is an isomorphism* *invariant of the two original actions. More precisely, any Borel conjugacy between G-actions gives a canonical homeomorphism between the spaces of orbit-joinings.*

Proof Consider the map

$$t : (f, p_1(z_1), p_2(z_2)) \to (f, p_1(z_1), r_f(p_2(z_2))).$$

This is a homeomorphism of $\mathscr{G} \times Z_1^G \times Z_2^G$, and carries any orbit joining $\hat{\mu}$ to a measure in $\mathscr{C}(\mu_1, \mu_2, \mathscr{G})$. As

$$t^{-1}(p_1(z_1), p_2(z_2), h) = (p_1(z_1), r_{f^{-1}}(p_2(z_2)), h),$$

$(t^{-1})^*$ is a homeomorphism of $\mathscr{C}(\mu_1, \mu_2, \mathscr{G})$ to its range. Those $\tilde{\mu} \in \mathscr{C}(\mu_1, \mu_2, \mathscr{G})$ with $(t^{-1})^*(\tilde{\mu})$ a $\hat{T}^{\alpha}$-invariant measure form a closed set homeomorphic (by $(t^{-1})^*$) to the space of orbit-joinings. If Φ_1 and Φ_2 are conjugacies of the original two actions to some other versions $(Z_i', \mathscr{F}_i', \mu_i', S_i^{\alpha_i'})$, then the map

$$(p_1(z_1), p_2(z_2), h) \to (p_1(\Phi_1(z_1)), p_2(\Phi_2(z_2)), h)$$

will give a t-equivariant map between their spaces of couplings, and hence (as in Corollary 6.5.3) a homeomorphism between their spaces of orbit-joinings. □

Corollary 6.5.6. *The space of C-orbit-joinings of two ergodic G-actions, as a topological space in the weak*-topology, is a conjugacy invariant, in that any pair of conjugacies of the joined actions gives a canonical homeomorphism between their spaces of C-orbit-joinings.*

Proof This is only slightly different from Corollary 6.5.5. Let Φ_1 and Φ_2 be a pair of conjugacies of the two joined actions to "primed" versions. Define a homeomorphism of $\vec{\mathscr{R}} \times Z_1^G \times Z_2^G$ by

$$t(p_1(z_1), p_2(z_2), \{f_i\}) = (p_1(z_1), r_{\ell(H(f_i))}(p_2(z_2)\{f_i\})).$$

For $\hat{\mu}$ a C-orbit joining, $t^*(\hat{\mu}) \in \mathscr{C}(\mu_1, \mu_2, \vec{\mathscr{R}})$. For the conjugacies Φ_1 and Φ_2 we can lift to the map

$$p_1(\Phi_1) \times p_2(\Phi_2) \times \mathrm{id} : Z_1^G \times Z_2^G \times \vec{\mathscr{R}} \to (Z_1')^G \times (Z_2')^G \times \vec{\mathscr{R}}$$

which will lift to a homeomorphism from $\mathscr{C}(\mu_1, \mu_2, \vec{\mathscr{R}})$ to $\mathscr{C}(\mu_1', \mu_2', \vec{\mathscr{R}})$ that is t^*-equivariant. Hence $(p_1(\Phi_1) \times p_2(\Phi_2) \times \mathrm{id})^*$ will be a homeomorphism between the spaces of C-orbit-joinings. □

Definition 6.5.7. *For ergodic and free G-actions* $(Z_1, \mathscr{F}_1, \mu_1, T_1^{\alpha_1})$ *and* $(Z_2, \mathscr{F}_2, \mu_2, T_2^{\alpha_2})$, *an* **m-joining** *of the two actions is a measure* $\hat{\mu} \in \mathscr{M}^m(Z_1 \times Z_2)$ *whose projections satisfy* $\pi_1^*(\hat{\mu}) = p_1^*(\mu_1)$ *and* $\pi_2^*(\overline{q}^*(\hat{\mu})) = p_2^*(\mu_2)$.

We write this space of measures as $J_m(T_1^{\alpha_1}, T_2^{\alpha_2})$, *suppressing the other components of the dynamical systems as long as they are obvious.*

Notice again that the space of m-joinings is not empty as it contains all ergodic joinings, extended via a point mass on $\{\mathrm{id}\}$ to a measure in $\mathscr{M}_{**}$. As $m(\alpha, \mathrm{id}) = 0$, this measure is in $\mathscr{M}^m(Z_1 \times Z_2)$. The following result follows directly from the definitions of the spaces involved.

Theorem 6.5.8. *The set* $J_m(T_1^{\alpha_1}, T_2^{\alpha_2})$ *is a Polish space as it is precisely the intersection of* $\mathscr{M}^m(Z_1 \times Z_2)$ *and the space of C-orbit-joinings of the two systems, both of which are Polish subspaces of* $\mathscr{M}_{**}(Z_1 \times Z_2)$. *As the topological space of C-orbit-joinings is a conjugacy invariant of the two actions joined, so too is the topological space of m-joinings.*

Proof Everything except perhaps the last remark is direct.

To clarify the last remark, in Lemmas 6.5.5 and 6.5.6 we saw that conjugacies of actions would lift to a canonical homeomorphism of their orbit-joinings and C-orbit-joinings. As this homeomorphism acts as the identity on the $\vec{\mathscr{R}}$-coordinate, it will preserve the subspace of m-joinings. □

Theorem 6.5.8 has been the goal of our entire journey from the beginning of Chapter 6. Knowing that the space of m-joinings is a Polish space now allows us to apply a category approach to the equivalence theorem. This is what we now proceed to do.

Notice that if $(\hat{Z}, \hat{\mathscr{F}}, \hat{\mu}, \hat{T}^{\alpha})$ is a free and ergodic G-action with full-group elements $\{\phi_i\}$ with

$$(f^{\alpha,\phi_1} \otimes f^{\alpha,\phi_2} \otimes \cdots)^* \mu \in \mathscr{M}^m$$

then of course there will be a β with $\alpha\phi_i \underset{m}{\rightarrow} \beta$. Now suppose we also have a pair of measure preserving maps $\pi_1 : \hat{Z} \to Z_1$ and $\pi_2 : \hat{Z} \to Z_2$ satisfying $\pi_1 \hat{T}^{\alpha} = T_1^{\alpha_1} \pi_2$ and $\pi_2 \hat{T}^{\beta} = T_2^{\alpha_2} \pi_2$. Then this G-action will give rise to an m-joining via the map $Q : \hat{Z} \to (Z_1 \times Z_2)^G \times \vec{\mathscr{R}}$ where

$$Q(\hat{z}) = (\{\pi_1(\hat{T}_g(\hat{z}))\}_{g \in G}, \{\pi_2(\hat{T}_g(\hat{z}))\}_{g \in G}, \{f^{\alpha,\phi_i}\})$$

as $Q^*(\hat{\mu}) \in J_m(T_1^{\alpha_1}, T_2^{\alpha_2})$.

We refer to such a G-action, sequence of full-group elements and projections π_1 and π_2 as an **m-joining in the loose sense** in that although it is not itself an element of $J_m(T_1^{\alpha_1}, J_2^{\alpha_2})$ it gives rise to one directly via the image $Q^*(\hat{\mu})$. The topology on $J_m(T_1^{\alpha_1}, T_2^{\alpha_2})$ puts a pseudotopology on these m-joinings in the loose sense.

Extending this one step further, if $\hat{T}^\alpha$ and $\hat{T}^\beta$ are two orbit-equivalent G-actions on $(\hat{Z}, \hat{\mathscr{F}}, \hat{\mu})$ and the maps $\pi_1 : \hat{Z} \to Z_1$ and $\pi_2 : \hat{Z} \to Z_2$ conjugate the action of $\hat{T}^\alpha$ to $T_1^{\alpha_1}$ and $\hat{T}^\beta$ to $T_2^{\alpha_2}$ respectively, then we refer to $(\hat{Z}, \hat{\mathscr{F}}, \hat{\mu})$, the pair of arrangements α and β and the factor maps π_1 and π_2 as an m-joining of $T_1^{\alpha_1}$ and $T_2^{\alpha_2}$ in the **very loose sense**. In this situation we can always interpolate full-group elements taking α to β rapidly enough to give an m-joining in the loose sense and hence an m-joining. This last notion of an m-joining in the very loose sense is the most natural one for what one would heuristically call an m-joining. The additional structure we have given is simply to place it in a Polish topological setting.

7
The Equivalence Theorem

7.1 Perturbing an m-equivalence

In proving the equivalence theorem, starting from one free and ergodic G-action, with two m-equivalent arrangements, we will need to construct a second one which will be, in a sense we now make precise, a perturbation of the first. The aim of this will be to show that a certain open set of m-joinings is in fact dense under appropriate hypotheses. This density will be shown by demonstrating how to "perturb" any given m-joining into the given open set. In this section we will develop the technical facts we will need for such perturbations.

Recall that when we omit the subscript x on an expression $f_x^{\alpha,\phi}$ or $h_x^{\alpha,\beta}$ we are regarding them as maps from X to $\mathscr{R}$ or $\mathscr{G}$. Suppose $(X, \mathscr{F}, \mu, T^\alpha)$ is a free and ergodic G-action and $\{\phi_i\}$ is a sequence of elements in the full-group of T^α for which

$$(f^{\alpha,\phi_1} \otimes f^{\alpha,\phi_2} \otimes \cdots)^* \mu \in \mathscr{M}^m.$$

The following discussion will be relative to this fixed action and sequence of full-group elements.

Under these conditions we know that $\alpha\phi_i \to \beta$ where $\alpha \overset{m}{\sim} \beta$. We describe now what we mean by an I_0, J, δ-perturbation of this G-action. It is perhaps more correct to call it an m-perturbation, but as we will assume m is a fixed size from here on, we omit this.

Throughout the rest of this chapter it will be convenient to have a metric for the weak*-topologies on various spaces of measures. Let D represent such a metric in all cases.

Definition 7.1.1. *Fix values I_0 and J and consider the measure*

$$\tilde{\mu} = (f^{\alpha,\phi_1} \otimes f^{\alpha,\phi_2} \otimes \cdots \otimes f^{\alpha,\phi_{I_0}} \otimes f^{\alpha,\phi_J})^* \mu \in \mathscr{M}_e(\mathscr{R}^{I_0+1}).$$

Suppose we have a second free and ergodic G-action $(X_1, \mathscr{F}_1, \mu_1, T_1^{\alpha_1})$ *with* I_0+1 *elements in its full-group* $\phi_1', \phi_1', \ldots, \phi_{I_0+1}'$, *and a further sequence of full-group elements* $\{\psi_i\}$ *satisfying:*

(i) $D(\tilde{\mu}, (f^{\alpha_1,\phi_1'} \otimes \cdots \otimes f^{\alpha_1,\phi_{I_0+1}'})^* \mu_1) < \delta$; *and*
(ii) $(f^{\alpha_1\phi_{I_0+1}',\psi_1} \otimes f^{\alpha_1\phi_{I_0+1}',\psi_2} \otimes \cdots)^* \mu_1 \in \mathscr{M}^m$; *and*
(iii) $m((f^{\alpha_1\phi_{I_0+1}',\psi_1} \otimes f^{\alpha_1\phi_{I_0+1}',\psi_2} \otimes \cdots)^* \mu_1) < \delta$.

We refer to such an action and collection of full-group elements as an **$\mathbf{I_0, J, \delta}$-perturbation** *of the original G-action and sequence in the full-group.*

Notice that in such an I_0, J, δ-perturbation the sequence of rearrangements $\alpha_1 \phi_{I_0+1} \psi_i$ will converge in i to an arrangement we will call β_1, and α_1 and β_1 will be m-equivalent. What interests us is the precise sequence of full-group elements

$$\phi_1', \phi_2', \ldots, \phi_{I_0+1}', \phi_{I_0+1}'\psi_1, \phi_{I_0+1}'\psi_2, \ldots$$

taking α_1 to β_1. We will write this sequence as $\{\phi_i''\}$.

Notice that if we drop to a subsequence of the ψ_i in an I_0, J, δ-perturbation, we will still have an I_0, J, δ-perturbation. What we will show in the rest of this section is that given a G-action $(X, \mathscr{F}, \mu, T^\alpha)$ and sequence of full-group elements $\{\phi_i\}$ with

$$(f^{\alpha,\phi_1} \otimes f^{\alpha,\phi_2} \otimes \cdots)^* \mu \in \mathscr{M}^m,$$

then for any I_0, if J is large enough and δ small enough, any I_0, J, δ-perturbation will still give

$$(f^{\alpha_1,\phi_1''} \otimes f^{\alpha_1,\phi_2''} \otimes \cdots)^* \mu_1 \in \mathscr{M}^m.$$

To begin to see why, notice the following calculation:
For all $i \leq I_0 + 1$,

$$m_{\alpha_1}(\phi_i'', \phi_j'') = \begin{cases} m_{\alpha_1}(\phi_i', \phi_j'), & j \leq I_0 + 1 \\ m_{\alpha_1}(\phi_i', \phi_{I_0+1}'\psi_{j-I_0-1}), & j > I_0 + 1 \end{cases}$$

$$\leq \begin{cases} m_{\alpha_1}(\phi_i', \phi_j'), & j \leq I_0 + 1 \\ m_{\alpha_1}(\phi_i', \phi_{I_0+1}') + m_{\alpha_1\phi_{I_0+1}'}(\mathrm{id}, \psi_{j-I_0-1}), & j > I_0 + 1 \end{cases}$$

$$\leq \begin{cases} m_{\alpha_1}(\phi_i', \phi_j'), & j \leq I_0 + 1 \\ m_{\alpha_1}(\phi_i', \phi_{I_0+1}') + \delta, & \text{for all } j > I_0 + 1 \text{ sufficiently large.} \end{cases}$$

Lemma 7.1.2. *For any free and ergodic G-action* $(X, \mathscr{F}, \mu, T^\alpha)$ *and full-group elements* ϕ_i *with*

$$(f^{\alpha,\phi_1} \otimes f^{\alpha,\phi_2} \otimes \cdots)^* \mu_1 \in \mathscr{M}^m$$

and value I_0, *there exists* $\delta > 0$ *such that for all J sufficiently large, any* I_0, J, δ*-perturbation (dropping to a subsequence of the* $\{\psi_i\}$ *if necessary) will satisfy*:

$$\limsup_{j\to\infty} m_{\alpha_1}(\phi_i'', \phi_j'') < 1/i$$

and

$$\limsup_{j\to\infty} m_{\beta_1}({\phi''}_i^{-1}, {\phi''}_j^{-1}) < 1/i.$$

Proof As $(f^{\alpha,\phi_1} \otimes \cdots)^* \mu \in \mathscr{M}^m$ we have, for all i,

$$\limsup_{j\to\infty} m_\alpha(\phi_i, \phi_j) < 1/i \quad \text{and}$$

$$\limsup_{j\to\infty} m_\beta(\phi_i^{-1}, \phi_j^{-1}) < 1/i.$$

Hence, for all i, both of these lim sup's are in fact limits converging to $m_\alpha(\phi_i, \langle\phi_j\rangle_\alpha)$ and $m_\beta(\phi_i^{-1}, \langle\phi_j^{-1}\rangle_\beta)$ both of which must be $< 1/i$. Hence there is an $\varepsilon_1 > 0$ so that for all $i \le I_0$ we have

$$m_\alpha(\phi_i, \langle\phi_j\rangle_\alpha) < 1/i - \varepsilon_1.$$

Thus for all J sufficiently large, and $i = 1, \ldots, I_0$,

$$m_\alpha(\phi_i, \phi_J) < 1/i - \varepsilon_1.$$

By Axiom 3, there exists $\delta > 0$ such that if

$$D(\tilde{\mu}, (f^{\alpha_1,\phi_1'} \otimes \cdots \otimes f^{\alpha_1,\phi_{I_0+1}'})^* \mu_1) < \delta$$

then all of the finite list of inequalities

$$m_{\alpha_1}(\phi_i', \phi_{I_0+1}') < 1/i - \varepsilon_1, \qquad i = 1, \ldots, I_0 + 1$$

will still hold.

As long as $\delta < \varepsilon_1/2$, the calculation made preceding this lemma implies

$$m_{\alpha_1}(\phi_i', \phi_{I_0+1}'\psi_j) \le m_{\alpha_1}(\phi_i', \phi_{I_0+1}') + m_{\alpha_1\phi_{I_0+1}'}(\mathrm{id}, \psi_j) < 1/i - \varepsilon_1/2$$

and for $i = 1, \ldots, I_0$

$$\limsup_{j\to\infty} m_{\alpha_1}({\phi''}_i, {\phi''}_j) < 1/i.$$

For $i = I_0 + 1$,

$$\limsup_{j\to\infty} m_{\alpha_1}(\phi''_{I_0+1}, \phi''_j) = m_{\alpha_1\phi''_{I_0+1}}(\mathrm{id}, \psi_{j-I_0-1}) < \delta.$$

Making sure that $\delta < 1/(I_0+1)$ we obtain this term. For $i > I_0 + 1$, as

$$\limsup_{j\to\infty} m_{\alpha_1}(\phi''_i, \phi''_j) = \limsup_{j\to\infty} m_{\alpha_1\phi''_{I_0+1}}(\psi_{i-I_0-1}, \psi_j)$$
$$= m_{\alpha_1\phi''_{I_0+1}}(\psi_{i-I_0-1}, \langle\psi_j\rangle_{\alpha_1}),$$

by dropping to a subsequence of the ψ_i we can ensure that this value is $< 1/i$.

As we assume that the sequence ϕ''_i achieves an m-equivalence between α_1 and β_1, the other set of strict inequalities now follows automatically from Lemma 6.4.5. □

We now want to obtain the same fact for pointwise convergence, that a small enough perturbation will not leave $\mathcal{M}_{**}$. The first step is to notice that from Axiom 2 we get the following:

Lemma 7.1.3. *For all $\delta > 0$ there exists $\delta_1 > 0$ and if $(X_1, \mathcal{F}_1, \mu_1, T_1^{\alpha_1})$ is a free and ergodic G-action, with ϕ'_{I_0+1} and $\{\psi_i\}$ in its full-group with $\langle\psi_i\rangle_{\alpha_1} \in \hat{E}_m(\alpha_1)$ and*

$$m_{\alpha_1\phi'_{I_0+1}}(\langle\psi_i\rangle_{\alpha_1}, \mathrm{id}) < \delta_1,$$

then for a subsequence of the ψ_i we have both

$$\int \sup_i(d(h^{\alpha_1\phi'_{I_0+1},\alpha_1\phi'_{I_0+1}\psi_i}, \mathrm{id}))\, d\mu_1 < \delta$$

and

$$\int \sup_i(d(h^{\beta_1,\beta_1\psi_i^{-1}}, \mathrm{id})) < \delta.$$

Proof Axiom 2 tells us that if δ_1 is sufficiently small then for all i sufficiently large,

$$\int d(h^{\alpha_1\phi'_{I_0+1},\alpha_1\phi'_{I_0+1}\psi_i}, \mathrm{id})\, d\mu_1 < \delta/4.$$

We can drop to a subsequence of the ψ_i with

$$d(h_x^{\alpha_1\phi'_{I_0+1},\alpha_1\phi'_{I_0+1}\psi_i}, h_x^{\alpha_1\phi'_{I_0+1},\beta_1}) \underset{i}{\to} 0$$

for μ_1-a.e. x. Hence by the Lebesgue dominated convergence theorem,

$$\int d(h^{\alpha_1\phi'_{I_0+1},\beta_1}, \mathrm{id}) \le \delta/4.$$

Thus by omitting sufficiently many initial terms ψ_i we can obtain

$$\sup_i (d(h_x^{\alpha_1\phi'_{I_0+1},\alpha_1\phi'_{I_0+1}\psi_i}, h_x^{\alpha_1\phi'_{I_0+1},\beta})) \le \delta/4$$

for a subset of X of measure $> 1 - \delta/4$.

As $d \le 1$, restricting to this tail of the ψ_i sequence we obtain

$$\int \sup_i (d(h^{\alpha_1\phi'_{I_0+1},\alpha_1\phi'_{I_0+1}\psi_i}, \mathrm{id}))\, d\mu_1 \le 3\delta/4 < \delta.$$

As the sequence $\{\psi_i\}$ realizes the m-equivalence between $\alpha_1\phi'_{I_0+1}$ and β_1, we also have

$$m_{\beta_1}(\langle\psi_i^{-1}\rangle_{\beta_1}, \mathrm{id}) < \delta_1$$

and the second inequality of the lemma follows symmetrically. □

Lemma 7.1.4. *Suppose $(X, \mathscr{F}, \mu, T^\alpha)$ is a free and ergodic G-action, such that $\{\phi_i\}$ is in its full-group with*

$$(f^{\alpha\phi_1} \otimes f^{\alpha\phi_2} \otimes \cdots)^*\mu \in \mathscr{M}^m.$$

Given any $\varepsilon > 0$ and I_0, there exists $\delta > 0$ and $J > I_0$ so that if $(X_1, \mathscr{F}_1, \mu_1, T_1^{\alpha_1})$, $\phi'_1, \dots, \phi'_{I_0+1}$ and $\{\psi_i\}$ form an I_0, J, δ-perturbation of the first system, then we can conclude:

(a) *for all $i \le I_0 + 1$*
$$\int R_{i,I_0+1}(h^{\alpha_1,\alpha_1\phi'_i}, \dots, h^{\alpha_1,\alpha_1\phi'_{I_0+1}})\, d\mu_1$$
$$\le \int R_{i,I_0+1}(h^{\alpha,\alpha\phi_i}, \dots, h^{\alpha,\alpha\phi_{I_0}}, h^{\alpha,\beta})\, d\mu + \varepsilon;$$

(b) $\int \sup_i d(h^{\alpha_1,\alpha_1\phi'_{I_0+1}\psi_i}, h^{\alpha_1,\alpha_1\phi'_{I_0+1}})\, d\mu_1 \le \varepsilon$;

(c) *for all $i \le I_0 + 1$,*
$$\int R_{i,I_0+1}(h^{\beta_1,\beta_1\phi'^{-1}_i}, h^{\beta_1,\beta_1\phi'^{-1}_{i+1}}, \dots, h^{\beta_1,\beta_1\phi'^{-1}_{I_0+1}})\, d\mu_1$$
$$\le \int R_{i,I_0+1}(h^{\beta,\beta\phi_i^{-1}}, h^{\beta,\beta\phi_{i+1}^{-1}}, \dots, h^{\beta,\beta\phi_{I_0}^{-1}}, h^{\beta,\alpha})\, d\mu + \varepsilon;$$
and

(d) $\int \sup_i d(h^{\beta_1,\beta_1\phi'^{-1}_{I_0+1}}, h^{\beta_1,\beta_1\psi_i^{-1}\phi'^{-1}_{I_0+1}})\, d\mu_1 < \varepsilon$.

Proof To begin, for any $\delta_1 > 0$ and I_0, there is a $\delta > 0$ such that for all J sufficiently large, perhaps dropping to a subsequence of the ψ_i, we will have:

(i) $D((f^{\alpha,\phi_1} \otimes \cdots \otimes f^{\alpha,\phi_{I_0}} \otimes h^{\alpha,\beta})^* \mu,$
$$(f^{\alpha_1,\phi'_1} \otimes \cdots \otimes f^{\alpha_1,\phi'_{I_0}} \otimes h^{\alpha_1,\alpha_1\phi'_{I_0+1}})^* \mu_1) < \delta_1;$$
(ii) $\int \sup_i (d(h^{\alpha_1\phi'_{I_0+1},\alpha_1\phi'_{I_0+1}\psi_i}, \mathrm{id}))\, d\mu_1 < \delta_1$; and
(iii) $\int \sup_i (d(h^{\beta_1,\beta_1\psi_i^{-1}}, \mathrm{id}))\, d\mu_1 < \delta_1$.

To obtain (i) just notice that $h_x^{\alpha,\alpha\phi_J} \underset{J}{\rightarrow} h_x^{\alpha,\beta}$ and the map H taking $\mathscr{R}$ to $\mathscr{G}$ is continuous. The second two statements follow directly from Lemma 7.1.3. The proofs of (a)–(d) will follow, except for one small step, from (i)–(iii) with δ_1 small enough and J large enough.

One can calculate that for any two G-arrangements α and β and full-group element ϕ that

$$f_x^{\beta,\phi^{-1}}(g) = h_x^{\alpha,\beta}(h_x^{\alpha\phi,\alpha}(h_x^{\beta,\alpha}(g) f_x^{\alpha,\phi}(\mathrm{id})))g^{-1}.$$

Represent a point in $\mathscr{R}^{I_0} \times \mathscr{G}$ as $(f'_1, f'_2, \ldots, f'_{I_0}, \ell')$, and define

$$F(f', \ell,)(g) = \ell'(H(f')^{-1}(\ell'^{-1}(g) f'(\mathrm{id})))g^{-1},$$

and notice that the maps

$$(f'_1, f'_2, \ldots, f'_{I_0}, \ell') \rightarrow H(F(f'_i, \ell')), \qquad i = 1, \ldots, I_0$$

are all continuous.

It follows that

$$H(F(f_{x_1}^{\alpha_1,\phi'_1}, h_{x_1}^{\alpha_1,\alpha_1\phi'_{I_0+1}})) = h_{x_1}^{\alpha_1\phi'_{I_0+1},\alpha_1\phi'_{I_0+1}\phi'^{-1}_i}.$$

Partition $\mathscr{R}^{I_0} \times \mathscr{G}$ into a countable collection of **clopen** sets according to the values

$$H(F(f'_i, \ell'))(g_j), H(f'_i)(g_j), \ell'(g_j) \quad \text{and}$$
$$H(F(f'_i, \ell'))^{-1}(g_j), H(f'_i)^{-1}(g_j), \ell'(g_j), \quad j = 1, \ldots, [\ln(8/\varepsilon)] + 1.$$

Label the partition elements $C_1, C_2, \ldots$. Notice that for any two points $(f'_1, \ldots, f'_{I_0}, \ell')$ and $(f''_1, \ldots, f''_{I_0}, \ell'')$ which belong to the same C_k we will have

$$d(\ell', \ell'') < \varepsilon/8$$
$$d(H(F(f'_i, \ell')), H(F(f''_i, \ell''))) < \varepsilon/8 \quad \text{and}$$
$$d(H(f'_i), H(f''_i)) < \varepsilon/8, \qquad i = 1, \ldots, I_0.$$

Choose a finite list $C_1, \dots, C_K$ with

$$\mu\big(\cup_{k=1}^{K}(f^{\alpha,\phi_1} \otimes \cdots \otimes f^{\alpha,\phi_{I_0}} \otimes h^{\alpha,\beta})^{-1}(C_k)\big) > 1 - \varepsilon/8.$$

As the C_k are clopen, we can choose δ_1 so small that (i) implies

$$(f^{\alpha_1,\phi'_1} \otimes \cdots \otimes f^{\alpha_1,\phi'_{I_0}} \otimes h^{\alpha_1,\alpha_1\phi'_{I_0+1}})^{*}\mu_1(\cup_{k=1}^{K} C_k) > 1 - \varepsilon/4.$$

Select an index $N > [\ln(8/\varepsilon)] + 1$ so large that for all g_j, $j = 1, \dots, [\ln(8/\varepsilon)] + 1$ and $(f'_1, \dots, f'_{I_0}, \ell') \in \cup_k C_k$ all of the values

$$H(f'_i)(g_j), H(f'_i)^{-1}(g_j), H(F(f'_i,\ell'))(g_j),$$
$$H(F(f'_i,\ell'))^{-1}(g_j), \ell'(g_j), \text{ and } \ell'^{-1}(g_j)$$

are indexed as some g_n, $n \le N$. (Remember, all of these values are constant on each set C_k.)

Set $e = 1/N$, be sure that

$$\delta_1 < \frac{e\varepsilon}{8(I_0+3)}$$

and (ii) now implies

$$\int d(h^{\alpha_1,\phi'_{I_0+1},\beta_1}, \mathrm{id})\, d\mu_1 \le \delta_1.$$

Define a set A by

$$A = \{x_1 : d(h_{x_1}^{\alpha_1\phi'_{I_0+1},\alpha_1\phi'_{I_0+1}\psi_i}, \mathrm{id}) \le e \quad\text{and}\quad d(h_{\phi'_j(x_1)}^{\alpha_1\phi'_{I_0+1},\alpha_1\phi'_{I_0+1}\psi_i}, \mathrm{id})$$
$$\le e \text{ for } j = 1, \dots, I_0 \text{ and all } i, \text{ hence } d(h_{x_1}^{\alpha_1\phi'_{I_0+1},\beta_1}, \mathrm{id})$$
$$\le e \text{ and } d(h_{\phi'_j(x_1)}^{\alpha_1\phi'_{I_0+1},\beta_1}, \mathrm{id}) \le e \text{ for } j = 1, \dots, I_0,$$

and we further require

$$d(h_{\phi'^{-1}_{I_0+1}(x_1)}^{\alpha_1\phi'_{I_0+1},\beta}, \mathrm{id}) \le e \quad\text{and}\quad d(h_{x_1}^{\beta_1,\beta_1\psi_i}, \mathrm{id}) \le e \text{ for all } i\}.$$

As there are $I_0 + 3$ inequalities to satisfy to lie in A, each of which holds on a set of measure at least δ_1/e (remember that μ_1 is preserved by all the ϕ'_i) we conclude that

$$\mu(A) > 1 - \varepsilon/8.$$

Let

$$A' = A \cap (f^{\alpha_1,\phi'_1} \otimes \cdots \otimes f^{\alpha_1,\phi'_{I_0}} \otimes f^{\alpha_1,\phi'_{I_0+1}})^{-1}(\cup_{k=1}^{K} C_k),$$

and $\mu_1(A') > 1 - 3\varepsilon/8$.

The four inequalities come in two pairs (a),(b) and (c),(d). The latter two are more difficult so we focus on their proofs. We begin with (c). For all $x_1 \in A'$, $i = 1, \dots, I_0$ and $j = 1, \dots, [\ln(8/\varepsilon)] + 1$ we will have

$$\begin{aligned} H(F(f_{x_1}^{\alpha_1,\phi'_i}, h_{x_1}^{\alpha_1,\beta_1}))(g_j) &= h_{x_1}^{\beta_1,\beta_1\phi'^{-1}_i}(g_j) \\ &= h_{x_1}^{\alpha_1\phi'_{I_0+1}\phi'^{-1}_i,\beta_1\phi'^{-1}_i} h_{x_1}^{\alpha_1\phi'_{I_0+1},\alpha_1\phi'_{I_0+1}\phi'^{-1}_i} h_{x_1}^{\beta_1,\alpha_1\phi'_{I_0+1}}(g_j) \\ &= h_{\phi_i(x_1)}^{\alpha_1\phi'_{I_0+1},\beta_1} h_{x_1}^{\alpha_1\phi'_{I_0+1},\alpha_1\phi'_{I_0+1}\phi'^{-1}_i} h_{x_1}^{\beta_1,\alpha_1\phi'_{I_0+1}}(g_j) \\ &= h_{x_1}^{\alpha_1\phi'_{I_0+1},\alpha_1\phi'_{I_0+1}\phi'^{-1}_i}(g_j) \end{aligned}$$

as both the pre- and post- functions in this composition act on g_j and its image as the identity. Taking the inverses of all the bijections in this calculation, we find the same identity holds there. That is to say:

$$h_{x_1}^{\beta\phi'^{-1}_i,\beta_1}(g_j) = h_{x_1}^{\alpha_1\phi'_{I_0+1}\phi'^{-1}_i,\alpha_1\phi'_{I_0+1}}(g_j).$$

Examining the $(I_0 + 1)$'th term and $j = 1, \dots, [\ln(8/\varepsilon)] + 1$,

$$\begin{aligned} h_{x_1}^{\beta_1,\beta_1\phi'^{-1}_{I_0+1}}(g_j) &= h_{x_1}^{\alpha_1,\beta_1\phi'^{-1}_{I_0+1}} h_{x_1}^{\alpha_1\phi'_{I_0+1},\alpha_1} h_{x_1}^{\beta_1,\alpha_1\phi'_{I_0+1}}(g_j) \\ &= h_{\phi'^{-1}_{I_0+1}(x_1)}^{\alpha_1\phi'_{I_0+1},\beta_1} h_{x_1}^{\alpha_1\phi'_{I_0+1},\alpha_1} h_{x_1}^{\beta_1,\alpha_1\phi'_{I_0+1}}(g_j) \\ &= h_{x_1}^{\alpha_1\phi'_{I_0+1},\alpha_1}(g_j), \end{aligned}$$

and again the same identity holds at these g_j for the inverse maps. Hence for $x_i \in A'$ and all $i = 1, \dots, I_0 + 1$,

$$d(h_{x_1}^{\beta_1,\beta_1\phi'^{-1}_i}, h_{x_1}^{\alpha_1\phi'_{I_0+1},\alpha_1\phi'_{I_0+1}\phi'^{-1}_i}) \le \varepsilon/8.$$

Calculating (c),

$$\begin{aligned} &\int R_{i,I_0+1}(h^{\beta_1,\beta_1\phi'^{-1}_i}, \dots, h^{\beta_1,\beta_1\phi'^{-1}_{I_0+1}})\, d\mu_1 \\ &\le \int_{A'} R_{i,I_0+1}(h^{\alpha_1\phi'_{I_0+1},\alpha_1\phi'_{I_0+1}\phi'^{-1}_i}, \dots, h^{\alpha_1\phi'_{I_0+1},\alpha_1})\, d\mu_1 + \frac{5\varepsilon}{8} \\ &\le \int R_{i,I_0+1}(h^{\alpha_1\phi'_{I_0+1},\alpha_1\phi'_{I_0+1}\phi'^{-1}_i}, \dots, h^{\alpha_1\phi'_{I_0+1},\alpha_1})\, d\mu_1 + \frac{5\varepsilon}{8}. \end{aligned}$$

The map

$$\begin{aligned} &(f'_1, f'_2, \dots, f'_{I_0+1}) \to \\ &\qquad R_{i,I_0}(H(F(f'_i, H(f'_{I_0+1}))), \dots, H(F(f'_{I_0+1}, H(f'_{I_0+1})))) \end{aligned}$$

is continuous from $\mathscr{R}^{I_0+1} \to [0, 1]$ and so if δ is small enough we will obtain from (i) of Definition 7.1.1 of an I_0, J, δ-perturbation that this calculation is

$$\leq \int R_{i,I_0+1}(h^{\alpha\phi_J,\alpha\phi_J\phi_i^{-1}}, \ldots, h^{\alpha\phi_J,\alpha})\, d\mu + \frac{3\varepsilon}{4}.$$

As $h_x^{\beta,\alpha\phi_J} \to \mathrm{id}$ for all x, if J now is sufficiently large, this is

$$\leq \int R_{i,I_0}(h^{\beta,\beta\phi_i^{-1}}, \ldots, h^{\beta,\beta\phi_{I_0+1}^{-1}}, h^{\beta,\alpha})\, d\mu + \varepsilon$$

which is (c).

To obtain (a) we can omit the first part of the argument for (c) and just notice that for $i = 1, \ldots, I_0 + 1$, the maps

$$(f'_1, \ldots, f'_{I_0+1}) \to R_{i,I_0+1}(H(f'_i), \ldots, H(f'_{I_0+1}))$$

taking $\mathscr{R}^{I_0+1} \to [0,1]$ are continuous and so from (i) of the definition of an I_0, J, δ-perturbation, if δ is small enough,

$$\begin{aligned} &\int R_{i,I_0+1}(h^{\alpha_1,\alpha_1\phi'_i}, \ldots, h^{\alpha_1,\alpha_1\phi'_{I_0+1}})\, d\mu_1 \\ &\qquad \leq \int R_{i,I_0+1}(h^{\alpha,\alpha\phi_i}, \ldots, h^{\alpha,\alpha\phi_{I_0}}, h^{\alpha,\alpha\phi_J}) + \frac{\varepsilon}{2} \end{aligned}$$

and now if J is chosen sufficiently large, this will be

$$\int R_{i,I_0}(h^{\alpha,\alpha\phi_i}, \ldots, h^{\alpha,\alpha\phi_{I_0}}, h^{\alpha,\beta})\, d\mu + \varepsilon$$

which is (a).

To demonstrate (d) notice that for $j = 1, \ldots, [\ln(8/\varepsilon)] + 1$, and $x_1 \in A'$ that

$$\begin{aligned} h_{x_1}^{\beta_1,\beta_1\phi'^{-1}_{I_0+1}}(g_j) &= h_{\phi'^{-1}_{I_0+1}(x_1)}^{\alpha_1\phi'_{I_0+1},\beta_1} h_{x_1}^{\alpha_1\phi'_{I_0+1},\alpha_1} h_{x_1}^{\beta_1,\alpha_1\phi'_{I_0+1}}(g_j) \\ &= h_{x_1}^{\alpha_1\phi'_{I_0+1},\alpha_1}(g_j) \end{aligned}$$

again as the pre- and post- composing functions act as the identity on g_j and its image. Also as before the same identity holds for the inverse bijections at g_j. We also have

$$h_{x_1}^{\beta_1,\beta_1\psi_i^{-1}\phi'^{-1}_{I_0+1}}(g_j) = h_{x_1}^{\beta_1,\beta_1\psi_i^{-1}} h_{\phi'^{-1}_{I_0+1}(x_1)}^{\alpha_1\phi'_{I_0+1},\beta_1} h_{x_1}^{\alpha_1\phi'_{I_0+1},\alpha_1} h_{x_1}^{\beta_1\psi_i^{-1},\alpha_1\phi'_{I_0+1}}(g_j).$$

For all x_1 we know that $h_{x_1}^{\beta_1\psi_i^{-1},\alpha_1\phi'_{I_0+1}}(g_j) = g_j$ once i is large enough.

Drop to a sufficiently distant tail of the sequence ψ_i so that on a set A'' with $\mu_1(A'') > 1 - \varepsilon/8$, for $x_1 \in A''$ we have

$$h_{x_1}^{\beta_1\psi_i^{-1},\alpha_1\phi'_{I_0+1}}(g_j) = g_j \text{ for all } j = 1,\ldots,N.$$

It follows that for $x_1 \in A' \cap A''$ we have

$$h_{x_1}^{\beta_1,\beta_1\psi_i^{-1}\phi'^{-1}_{I_0+1}}(g_j) = h_{x_1}^{\alpha_1\phi'_{I_0+1},\alpha_1}(g_j),$$

and the same identity holds on the inverses of these bijections at g_j. (Note: it is this last remark that required us to use N in the definition of A'' and not $[\ln(8/\varepsilon)] + 1$.)

This implies for all $x_1 \in A' \cap A''$, for all i and for $1 \le j \le [\ln(8/\varepsilon)] + 1$ that

$$h_{x_1}^{\beta_1,\beta_1\phi'^{-1}_{I_0+1}}(g_j) = h_{x_1}^{\beta_1,\beta_1\psi_i^{-1}\phi'^{-1}_{I_0+1}}(g_j)$$

and the same identity holds for the inverses of these bijections at g_j. Hence

$$\int \sup_i d(h^{\beta_1,\beta_1\phi'^{-1}_{I_0+1}}, h^{\beta_1,\beta_1\psi_i^{-1}\phi'^{-1}_{I_0+1}})\, d\mu_1$$

$$\le \int_{A'\cap A'} \sup_i d(h^{\beta_1,\beta_1\phi'^{-1}_{I_0+1}}, h^{\beta_1,\beta_1\psi_i^{-1}\phi'^{-1}_{I_0+1}})\, d\mu_1 + \mu_1(A'^c \cup A''^c)$$

$$\le \frac{\varepsilon}{4} + \frac{\varepsilon}{8} < \varepsilon.$$

For (b) we follow similar but easier lines. Notice that for $j = 1,\ldots,[\ln(8/\varepsilon)] + 1$ and $x_1 \in A'$ that

$$h_{x_1}^{\alpha_1,\alpha_1\phi'_{I_0+1}\psi_i}(g_j) = h_{x_1}^{\alpha_1\phi'_{I_0+1},\alpha_1\phi'_{I_0+1}\psi_i} h_{x_1}^{\alpha_1,\alpha_1\phi'_{I_0+1}}(g_j) = h_{x_1}^{\alpha_1,\alpha_1\phi'_{I_0+1}}(g_j)$$

and the same identity holds at g_j for the inverse bijections. Now integrate just as for (d) to obtain (b). □

Theorem 7.1.5. *For $(X, \mathscr{F}, \mu, T^\alpha)$ a free and ergodic G-action and full-group elements ϕ_i with*

$$(f^{\alpha,\phi_1} \otimes f^{\alpha,\phi_1} \otimes \cdots)^*\mu \in \mathscr{M}^m$$

and I_0, there is a $\delta > 0$ such that for all J sufficiently large, any I_0, J, δ-perturbation

$$(X_1, \mathscr{F}_1, \mu_1, T_1^{\alpha_1}), \phi'_1, \ldots \phi'_{I_0+1}, \text{ and } \{\psi_i\},$$

(by dropping to a subsequence of the ψ_i) will satisfy

$$(f^{\alpha_1,\phi''_1} \otimes f^{\alpha_1,\phi''_2} \otimes \cdots)^* \mu_1 \in \mathcal{M}^m$$

as well.

Proof Having already proven in Lemma 7.1.2 that if this measure is in $\mathcal{M}_{**}$ it will be in $\mathcal{M}^m$ all we need to show is

(a) $\int R_i(h^{\alpha_1,\alpha_1\phi''_i}, h^{\alpha_1,\alpha_1\phi''_{i+1}}, \dots)\, d\mu_1 < (2^{i+2}i)^{-1}$ and

(b) $\int R_i(h^{\beta_1,\beta_1\phi''^{-1}_i}, h^{\beta_1,\beta_1\phi''^{-1}_{i+1}}, \dots)\, d\mu_1 < (2^{i+2}i)^{-1}$.

As $(f^{\alpha,\phi_1} \otimes f^{\alpha,\phi_1} \otimes \cdots)^* \mu_1 \in \mathcal{M}^m$ we already know that there must exist an $\varepsilon_1 > 0$ so that for all $i = 1, \dots, I_0$,

$$\int R_i(h^{\alpha,\alpha\phi_i}, \dots)\, d\mu < (2^{i+2}i)^{-1} - \varepsilon_1 \quad \text{and}$$

$$\int R_i(h^{\beta,\beta\phi_i^{-1}}, \dots)\, d\mu < (2^{i+2}i)^{-1} - \varepsilon_1.$$

To demonstrate (a) for $i = 1, \dots, I_0$, notice

$$\int R_i(h^{\alpha_1,\alpha_1\phi''_i}, \dots)\, d\mu_1 \le \int R_{i,I_0+1}(h^{\alpha_1,\phi'_i}, \dots, h^{\alpha_1,\alpha_1\phi'_{I_0+1}})\, d\mu_1$$
$$+ \int \sup_j d(h^{\alpha_1,\alpha_1\phi'_{I_0+1}}, h^{\alpha_1,\alpha_1\phi'_{I_0+1}\psi_j})\, d\mu_1.$$

Applying Lemma 7.1.4 with "ε" $= \varepsilon_1/3$, make sure δ is at most the "δ" obtained there, and for all J sufficiently large we obtain from (a) and (b) of Lemma 7.1.4 that this calculation is

$$\le \int R_{i,I_0+1}(h^{\alpha,\alpha\phi_i}, \dots, h^{\alpha,\alpha\phi_{I_0}}, h^{\alpha,\beta})\, d\mu + 2\varepsilon_1/3$$
$$\le \int R_i(h^{\alpha,\phi_i}, h^{\alpha,\phi_{i+1}}, \dots)\, d\mu$$

(as $h^{\alpha,\alpha\phi_i} \to h^{\alpha,\beta}$)

$$\le (2^{i+1}i)^{-1} - \varepsilon_1/3 < (2^{i+2}i)^{-1}.$$

For $i \ge I_0 + 2$,

$$\int R_i(h^{\alpha_1,\alpha_1\phi''_i}, \dots)\, d\mu_1 = \int R_i(h^{\alpha_1,\alpha_1\phi'_{I_0+1}\psi_{i-I_0-1}}, \dots)\, d\mu_1$$

and as $\alpha_1\phi_{I_0+1}\psi_{i-I_0-1} \underset{i}{\to} \beta_1$, by dropping to a subsequence of the ψ_i we can ensure this is

$$\leq \int \sup_j (2d(h^{\alpha_1,\alpha_1\phi'_{I_0+1}\psi_{i-I_0-1}}, h^{\alpha_1,\beta_1}))\, d\mu_1$$
$$< (2^{i+2}i)^{-1}.$$

This completes (a).

Obtaining (b) follows parallel lines using (c) and (d) of Lemma 7.1.4. For $i = 1,\ldots,I_0+1$,

$$\int R_i(h^{\beta_1,\beta_1\phi''^{-1}_i},\ldots)\, d\mu_1 \leq \int R_{i,I_0+1}(h^{\beta_1,\beta_1\phi'^{-1}_i},\ldots,h^{\beta_1,\beta_1\phi'^{-1}_{I_0+1}})\, d\mu_1$$
$$+ \int \sup_i d(h^{\beta_1,\beta_1\phi'^{-1}_{I_0+1}}, h^{\beta_1,\beta_1\psi_i^{-1}\phi'^{-1}_{I_0+1}})\, d\mu_1$$
$$\leq \int R_i(h^{\beta,\beta\phi_i^{-1}},\ldots)\, d\mu + 2\varepsilon_1/3$$

from (c) and (d) of Lemma 7.1.4 and that $\beta\phi_i^{-1} \to \alpha$. But this is

$$\leq (2^{i+1}i)^{-1} - \varepsilon_1/3 < (2^{i+2}i)^{-1}.$$

For $i \geq I_0 + 2$ we calculate

$$\int R_i(h^{\beta_1,\beta_1\phi''^{-1}_i},\ldots)\, d\mu_1 = \int R_i(h^{\beta_1,\beta_1\psi^{-1}_{i-I_0-1}\phi'^{-1}_{I_0+1}},\ldots)\, d\mu_1$$

and since $\beta_1\psi^{-1}_{i-I_0-1}\phi'^{-1}_{I_0+1} \underset{i}{\to} \alpha_1$, by dropping to a subsequence of the ψ_i, just as for the previous case, we can ensure this is

$$< (2^{i+2}i)^{-1}.$$

□

7.2 The $\overline{m}$-distance and m-finitely determined processes

Definition 7.2.1. *Suppose $(Z,\mathscr{F},\mu,T^\alpha)$ is a free and ergodic G-action and $P : Z \to \Sigma$ is a finite labeled partition (that is to say, Σ is a finite labeling set). We refer to $(Z,\mathscr{F},\mu,T^\alpha,P)$, as a Σ-valued process. We will usually abbreviate this by just the pair (T^α,P) as long as there is no confusion. We consider two Σ-valued processes to be identical if they give rise to the same σ-invariant measure on Σ^G via the usual map of a point in Z to its name $\{P(T^\alpha_g(z))\}_{g\in G} \in \Sigma^G$.*

If it is not necessary, we will also suppress the Σ *and just refer to* (T^{α}, P) *as a process. This is in keeping with the usual vocabulary of the isomorphism theory of Ornstein and Weiss.*

Definition 7.2.2. *Given two* Σ*-processes* $(T_1^{\alpha_1}, P_1)$ *and* $(T_2^{\alpha_2}, P_2)$, *we define the* **$\overline{\mathbf{m}}$-distance** *between them by*

$$\overline{m}(T_1^{\alpha_1}, P_1; T_2^{\alpha_2}, P_2) = \inf_{\hat{\mu} \in J_m(T_1^{\alpha_1}, T_2^{\alpha_2})} \big(\hat{\mu}(\{(\vec{z}_1, \vec{z}_2) : P_1(z_{1,\mathrm{id}}) \neq P_2(z_{2,\mathrm{id}})\}) + m(\hat{\mu})\big).$$

That is to say, we consider all m-joinings of the two actions, and among them look for the one which simultaneously matches the two labeled partitions on the two processes as closely as possible, and is as small an m-joining as possible.

For $\hat{\mu}$ an m-joining of two processes we abbreviate

$$\hat{\mu}(\{(\vec{z}_1, \vec{z}_2) : P_1(z_{1,\mathrm{id}}) \neq P_2(z_{2,\mathrm{id}})\}) \text{ by } \hat{\mu}(P_1 \triangle P_2).$$

We refer to the evaluation $\hat{\mu}(P_1 \triangle P_2) + m(\hat{\mu})$ as the **$\overline{\mathbf{m}}$-evaluation** at $\hat{\mu}$.

Lemma 7.2.3. *The* $\overline{m}$*-distance is a metric on the space of all* Σ*-valued processes. Furthermore, for any* $\varepsilon > 0$, *the set of m-joinings* $\hat{\mu}$ *with*

$$\hat{\mu}(P_1 \triangle P_2) + m(\hat{\mu}) < \overline{m}(T_1^{\alpha_1}, P_1; T_2^{\alpha_2}, P_2) + \varepsilon$$

is open.

Proof That $\overline{m}$ is symmetric follows from the fact that $\overline{q}^*$ interchanges the roles of the two processes in an m-joining but does not alter the calculation of the infimum.

The existence of the diagonal joining certainly tells us that the $\overline{m}$-distance between a process and itself is zero. For the other direction of this, suppose $\overline{m}(T_1^{\alpha_1}, P_1; T_2^{\alpha_2}, P_2) = 0$. Let $\hat{\mu}_i$ be a sequence of m-joinings of the two actions with the $\overline{m}$-evaluation converging to zero. Consider the projections of these measures to their $\Sigma^G \times \Sigma^G$-names alone, dropping all other coordinates. Certainly then, the measures must be converging to a measure supported on the diagonal. On the first coordinate it must be $p_1^*(\mu_1)$, the projection of μ_1 onto the space of Σ-names. As $m(\hat{\mu}_i) \to 0$, by Axiom 2 we must also have that the projections of the $\hat{\mu}_i$ to measures on $\mathscr{G}$ are converging to a point mass on the identity bijection. That is to say, the $\hat{\mu}_i$ converge on $\Sigma^G \times \Sigma^G \times \mathscr{G}$ to a joining (not just an orbit-joining) of the two processes. As it is supported on the diagonal, the two processes must be identical. For the triangle inequality, notice that if we have two m-joinings, $\hat{\mu}_1$ of $(T_1^{\alpha_1}, P_1)$ and $(T_2^{\alpha_2}, P_2)$ and $\hat{\mu}_2$ of $(T_1^{\alpha_1}, P_1)$ with $(T_3^{\alpha_3}, P_3)$, then we can construct the relatively independent coupling of $\hat{\mu}_1$ and $\hat{\mu}_2$ over their common $(T_1^{\alpha_1}, P_1)$-factor. An ergodic component of this $(\hat{\mu}_3)$ will be an ergodic and free action with three m-equivalent arrangements, (versions of α_1, α_2 and α_3), and three Σ-valued partitions (versions of P_1, P_2 and P_3). In this fixed system, $m_{\hat{\mu}_3,\alpha_1}$ is a metric on its m-equivalence class of arrangements, and the calculation $\hat{\mu}_3(P_i, P_j)$ is a metric on Σ-valued partitions.

Projecting $\hat{\mu}_3$ to just its $Z_2 \times Z_3$-coordinates and partitions, $\hat{\mu}_3$ can be mapped directly to an m-joining of the two processes $(T_2^{\alpha_2}, P_2)$ and $(T_3^{\alpha_3}, P_3)$. The $\overline{m}$-evaluation for this m-joining can be pulled back and evaluated on $\hat{\mu}_3$ where it will be maximized by the sum of the two $\overline{m}$-calculations for the original two m-joinings $\hat{\mu}_1$ and $\hat{\mu}_2$.

That the set of m-joinings that get within ε of the $\overline{m}$-distance is open follows from an observation. Knowing that the space of m-joinings is a conjugacy invariant means we can assume here that the maps P_1 and P_2 are continuous (i.e. the partitions are into clopen sets) as this is simply a different choice of model. This means the $\overline{m}$-evaluation is an upper semi-continuous function of the m-joining $\hat{\mu}$. □

We now define the notion of an m-finitely determined process. To readers familiar with the isomorphism theory this is lifted directly from the corresponding notion there. Hence it will be automatic that for m an entropy-preserving size, the Bernoulli processes are m-finitely determined, giving a basic class of examples. In this context it is more natural to speak of m-finitely determined processes of zero m-entropy and of positive m-entropy. The Bernoulli processes provide examples of m-finitely determined processes of positive m-entropy. One must examine case by case whether m-finitely determined processes of zero m-entropy exist. The simplest example is, of course, that no free and ergodic finitely determined actions of zero entropy exist.

Having defined the m-finitely determined property, we will first show that it is an m-equivalence invariant. We will show more, that in fact any process that sits as a factor of an m-finitely determined process is again m-finitely determined.

Hence one can speak of an m-finitely determined G-action as one for which every partition is m-finitely determined.

Finally we will show that among the m-finitely determined actions, m-entropy is a complete invariant of m-equivalence, i.e. that any two m-

finitely determined actions of the same m-entropy are in fact m-equivalent. We do this by showing that if $(X, \mathscr{F}, \mu, T^\alpha)$ is m-finitely determined, and $(X_1, \mathscr{F}_1, \mu_1, T_1^{\alpha_1})$ is any other free and ergodic G-action with $h_m(T_1^{\alpha_1}) \geq h_m(T^\alpha)$, then in the space of m-joinings $J_m(T^\alpha, T_1^{\alpha_1})$, those $\hat{\mu}$ for which the full-group elements $\{\phi_i\}$ and the first coordinate algebra $\mathscr{F}_1$ are $\hat{\mu}$-a.s. $\mathscr{F}_1$-measurable, form a dense G_δ subset. If the two actions happened to both be m-finitely determined and of the same m-entropy, then, simply intersecting the two residual subsets, we see that m-equivalences between them not only exist but form a dense G_δ subset of their space of m-joinings.

Before stating the definition of m-finitely determined, following our earlier convention, fix a metric D giving the weak*-topology on the space of Borel measures on a sequence space Σ^G where Σ is some finite labeling space. Also remember, for $P : X \to \Sigma$, we let $p(x) = \{P(T_g^\alpha(x))\}_{g \in G}$ be the T^α, P-name of the point. Again as a reminder, we define a pseudometric

$$\|(T, P), (T_1, P_1)\|_* = D(p^*(\mu), p_1^*(\mu_1)).$$

Definition 7.2.4. *We say a Σ-valued process (T^α, P) is* **m-finitely determined** *(abbreviated m-f.d.) if for any $\varepsilon > 0$ there is a $\delta > 0$ so that if $(T_1^{\alpha_1}, P_1)$ is any other Σ-valued process satisfying:*

(1) $\|(T^\alpha, P), (T_1^{\alpha_1}, P_1)\|_* < \delta$ *and*

(2) $h_m(T_1^{\alpha_1}, P_1) \geq h_m(T^\alpha, P) - \delta$,

then

(3) $\overline{m}(T_1^{\alpha_1}, P_1; T^\alpha, P) < \varepsilon$.

A few comments are appropriate. To begin, condition (2) is a bit disingenuous in that if m is entropy-preserving, then it should read

$$h(T_1^{\alpha_1}, P_1) \geq h(T^\alpha, P) - \delta$$

and if m is entropy-free, then (2) is no condition at all.

We have mentioned earlier that among positive m-entropy processes (that is to say, positive entropy processes and entropy-preserving sizes m), the Bernoulli processes give examples of m-f.d. processes. This is simply because for the f.d. processes of Ornstein and Weiss (1) and (2) imply $\bar{d}$-closeness, which is to say, the existence of a joining $\hat{\mu}$ with

$$\hat{\mu}(P \triangle P_1) < \varepsilon.$$

Such a joining of course extends via a point mass on $\{\text{id}\}$ to an m-joining with $m(\hat{\mu}) = 0$.

Although this definition is well-suited to the completion of a general equivalence theorem it suffers from being perhaps difficult to verify for any particular process. For 3^+ sizes we can use a more easily verified condition. In its definition we use the notion of an **ergodic lift** of some G-action T on a standard space $(X, \mathscr{F}, \mu)$. By this we mean an ergodic G-action $\hat{T}$ on some $(\hat{X}, \hat{\mathscr{F}}, \hat{\mu})$ which factors by some map π onto T. Notice that the full-group of T then lifts as a subgroup of the full-group of $\hat{T}$. Further, if $T = T^\alpha$ then any other arrangment β of the orbit of T lifts as well. Both of these observations follow from the fact that all functions on X, in particular $f^{\alpha,\phi}$, $q^{\alpha,\phi}$ and $h^{\alpha,\beta}$ lift to $\hat{X}$ through the projection of $\hat{x}$ to $\pi(\hat{x})$. The same of course applies to say that any partition P of X can be regarded as a partition of $\hat{X}$.

Definition 7.2.5. *For T^α a free and ergodic G-action we say the Σ valued process (T^α, P) is* **weakly m-finitely determined** *(weakly m-f.d.) if for each $\varepsilon > 0$ there is a $\delta > 0$ such that for any other free and ergodic G-action $T_0^{\alpha_0}$ and Σ valued partition P_0 with*

(1) $\|(T^\alpha, P), (T_0^{\alpha_1}, P_0)\|_* < \delta$ *and*

(2) $h_m(T_0^{\alpha_0}, P_0) \geq h_m(T^\alpha, P) - \delta$

and each $\delta_1 > 0$ there is an ergodic lift $T_1^{\alpha_1}$ of $T_0^{\alpha_0}$, a Σ-valued partition P_1 of X_1 and a ϕ_1 in the full-group of $T_1^{\alpha_1}$ with

(a) $\mu_1(P_0 \triangle P_1) < \varepsilon$ *and*

(b) $m(\alpha_1, \phi_1) < \varepsilon$,

for which

(1′) $\|(T^\alpha, P), (T_1^{\alpha_1 \phi_1}, P_1)\|_* < \delta_1$, *and*

(2′) $h_m(T_1^{\alpha_1}, P_1) \geq h_m(T^\alpha, P) - \delta_1$.

Notice that this condition becomes particularly simple for an entropy-free size where both conditions (2) and (2′) are vacuous.

Theorem 7.2.6. *Suppose T^α is an ergodic and free action of G and P is a Σ-valued partition (Σ a finite set).*

(i) *For each size m if (T^α, P) is m-finitely determined then it is weakly m-finitely determined.*

(ii) *For each 3^+ size m if (T^α, P) is weakly m-finitely determined then it is m-finitely determined.*

Proof For Part (i). suppose (T^α, P) is m-finitely determined. If δ is small enough and $(T_0^{\alpha_0}, P_0)$ satisfies (1) and (2) of m-f.d. then we obtain

$$\overline{m}(T_0^{\alpha_0}, P_0; T^\alpha, P) < \varepsilon$$

and there is a $\hat{\mu} \in J_m(T_0^{\alpha_0}, T^{\alpha})$ with

$$\hat{\mu}(\{(\vec{z}_1, \vec{z}_2) : P_1(z_{1,\text{id}} \neq P_2(z_{2,\text{id}})\}) + m(\hat{\mu}) < \varepsilon.$$

The action $\sigma^{\boldsymbol{\alpha}}$ on this m-joining is an ergodic lift of $T_0^{\alpha_0}$ and

$$m_{\hat{\mu}}(\boldsymbol{\alpha}, \boldsymbol{\beta}) < \varepsilon$$

which is to say

$$\lim_{i\to\infty} m_{\hat{\mu}}(\boldsymbol{\alpha}, \boldsymbol{\alpha\phi}_i) < \varepsilon.$$

As the sequence $\boldsymbol{\alpha\phi}_i$ is converging in $m_{\hat{\mu}}$ to $\boldsymbol{\beta}$ we have

$$(\sigma^{\boldsymbol{\alpha\phi}_i}, P_2(z_{2,\text{id}})) \longrightarrow (\sigma^{\boldsymbol{\beta}}, P_2(z_{2,\text{id}}))$$

in both distribution and m-entropy. Hence for this lift of $T_0^{\alpha_0}$ we can select $\phi = \boldsymbol{\phi}_i$ for some i and $P_1 = P_2(z_{2,\text{id}})$ and obtain (a), (b), (1′) and (2′).

To verify (ii) notice that the weakly m-f.d. condition is set to be used inductively moving from conditions (1) and (2) to (1′) and (2′) with (a) and (b) measuring the size of successive perturbations. Thus if (T^{α}, P) is weakly m-f.d. and $\varepsilon > 0$, then for any $(T_0^{\alpha_0}, P_0)$ satisfying (1) and (2) of the definition one can obtain a succession of ergodic lifts, partitions P_i and full-group elements ϕ_i. To simplify notation we can assume all these ergodic lifts sit inside one maximal ergodic lift we call $\hat{T}_0^{\hat{\alpha}_0}$. From such an inductive application of the definition we obtain for an initial step:

$$m(\hat{\alpha}_0, \phi_1) < \varepsilon \text{ and}$$
$$\hat{\mu}(P_0 \triangle P_1) < \varepsilon$$

but after this first step, for $i \geq 1$

$$m(\hat{\alpha}_0\phi_i, \phi_i^{-1}\phi_{i+1}) < \varepsilon_i \text{ and}$$
$$\hat{\mu}(P_i \triangle P_{i+1}) < \varepsilon_i$$

where the values ε_i are at our disposal to choose.

Hence we can force

$$\hat{\alpha}_i \xrightarrow{m} \hat{\beta}$$

with $m(\hat{\alpha}_0, \hat{\beta}) < \varepsilon$ and

$$P_i \longrightarrow \hat{P}$$

with $\hat{\mu}(P_0 \triangle \hat{P}) < \varepsilon$.

As $(\hat{T}_0^{\hat{\alpha}_0\phi_i}, P_i) \longrightarrow (T^\alpha, P)$ in distributions we will have

$$(\hat{T}_0^{\hat{\beta}}, \hat{P}) \equiv (T^\alpha, P).$$

If m is a 3^+ size then Theorem 2.2.7 ensures $\hat{\beta}\phi_i^{-1} \xrightarrow{m} \hat{\alpha}_0$ and that $\hat{\alpha}_0$, $\hat{\beta}$ and the sequence ϕ_i form an m-joining in the weak sense of $T_0^{\alpha_0}$ and T^α allowing us to conclude that

$$\overline{m}(T_0^{\alpha_0}, P_0; T^\alpha, P) < \varepsilon.$$

□

Our next goal is to see that the m-f.d. property is actually a property of a sub-σ-algebra, not just a partition. More precisely, for any process (T^α, P), let $\mathscr{H}(T^\alpha, P)$ be the σ-algebra

$$\bigvee_{g \in G} T_g^\alpha(P),$$

the smallest T^α-invariant σ-algebra relative to which P is measurable. What we want to demonstrate is the following.

Theorem 7.2.7. *If (T^α, P) is m-f.d., and $Q : X \to \Sigma_Q$ is any finite $\mathscr{H}(T^\alpha, P)$-measurable partition, then (T^α, Q) is m-f.d.*

We will argue this assuming m is entropy-preserving. The entropy-free case follows precisely the same lines, but without the need for any entropy estimates. The result will follow from the next two lemmas.

Lemma 7.2.8. *Suppose (T^α, P) is m-f.d. and $Q : X \to \Sigma_Q$ is another finite partition with*

$$\mathscr{H}(T^\alpha, P) = \mathscr{H}(T^\alpha, Q).$$

Then (T^α, Q) is m-f.d.

Proof We prove this for m an entropy-preserving size. Fix $\varepsilon > 0$ and choose $K_0 \subseteq G$, a finite subset, so that there is a "coding"

$$c_0 : \Sigma_P^{K_0} \to \Sigma_Q$$

satisfying

$$\mu(\{x : c_0(\{P(T_g^\alpha(x))\}_{g \in K_0}) \neq Q(x)\}) < \varepsilon/10.$$

Choose $\varepsilon_0 < \frac{\varepsilon}{10\#K_0}$ small enough so that on any orbit space for any pair of m-equivalent G-arrangements α_1 and α_2 with $m(\alpha_1, \alpha_2) < \varepsilon_0$, by Axiom 2 we will have

$$h_x^{\alpha_1,\alpha_2}|_{K_0} = \mathrm{id}$$

for all but at most $\varepsilon/10$ in measure of the points x.

Use ε_0 in the definition of m-f.d. for the process (T^α, P) to obtain a value δ_0. Choose $\bar{\varepsilon} < \delta_0/10$ with

$$\bar{\varepsilon}\log(\#\Sigma_Q) + H(\bar{\varepsilon}) < \delta_0/10$$

and

$$\bar{\varepsilon} < \frac{\varepsilon_0}{10\#K_0}.$$

Now once more choose finite codes c_1 and c_2 (much more accurate than c_0)

$$c_1 : (\Sigma_Q)^K \to \Sigma_P \text{ and } c_2 : (\Sigma_P)^K \to \Sigma_Q,$$

K a finite subset of G, satisfying

(a) $\mu(\{x : c_1(\{Q(T_g^\alpha(x))\}_{g\in K}) = P(x)\}) > 1 - \bar{\varepsilon}$,

and letting $\hat{P}(x) = c_1(\{Q(T_g^\alpha(x))\}_{g\in K})$,

(b) $\mu(\{x : c_2(\{\hat{P}(T_g^\alpha(x))\}_{g\in K}) = c_2(\{P(T_g^\alpha(x))\}_{g\in K}) = Q(x)\})$

$> 1 - \bar{\varepsilon}$.

For any Σ_Q-valued process $(T_1^{\alpha_1}, Q_1)$ let

$$\hat{P}_1(x_1) = c_1(\{Q_1(T_{1,g}^{\alpha_1}(x_1)\}_{g\in K}).$$

Choose $\delta > 0$ so that if

(1) $\|(T, Q), (T_1, Q_1)\|_* < \delta$ then

(1′) $\|(T^\alpha, P), (T_1^{\alpha_1}, \hat{P}_1)\|_* \leq \|(T^\alpha, \hat{P}), (T^\alpha, P)\|_*$
$+\|(T^\alpha, \hat{P}), (T_1^{\alpha_1}, \hat{P}_1)\|_* < \delta_0,$

and furthermore both

$$\mu_1(\{x_1 : c_0(\{\hat{P}_1(T_{1,g}^{\alpha_1}(x_1))\}_{g\in K}) = Q_1(x_1)\}) > 1 - \varepsilon/5$$

and

$$\mu_1(\{x_1 : c_2(\{\hat{P}_1(T_{1,g}^{\alpha_1}(x_1))\}_{g\in K}) = Q_1(x_1)\}) > 1 - 2\bar{\varepsilon}.$$

Hence

$$h(T_1^{\alpha_1}, \hat{P}_1) \geq h(T_1^{\alpha_1}, Q_1) - 2\bar{\varepsilon}\log(\#\Sigma_Q) - H(2\bar{\varepsilon})$$
$$> h(T_1^{\alpha_1}, Q_1) - \delta/10.$$

Thus if

(2) $h(T_1^{\alpha_1}, Q_1) > h(T^\alpha, Q) - \delta = h(T^\alpha, P) - \delta$ then

(2′) $h(T_1^{\alpha_1}, \hat{P}_1) > h(T^\alpha, P) - \delta_0$, and hence

(3′) $\overline{m}(T_1^{\alpha_1}, \hat{P}_1; T^\alpha, P) < \varepsilon_0$.

Let $\hat{\mu}$ be an ergodic joining of T^α and $T_1^{\alpha_1}$ for which the $\overline{m}$-evaluation is less than ε_0, that is to say, for which

$$\hat{\mu}(\hat{P}_1 \triangle P) < \varepsilon_0 \text{ and } m(\hat{\mu}) < \varepsilon_0.$$

We conclude that

$\hat{\mu}(Q_1 \triangle Q) < \hat{\mu}(\{(p_1(x_1), p(x), \{f_i\}) :$ either

(i) $\ell(H(f_i))(g) \neq g$ for some $g \in K_0$,

(ii) $\hat{P}_1(T_{1,g}^{\alpha_1}(x_1)) \neq P(T_g^\alpha(x))$ for some $g \in K_0$,

(iii) $c_0(\{\hat{P}_1(T_{1,g}^{\alpha_1}(x_1))\}_{g \in K_0}) \neq Q_1(x_1)$ or,

(iv) $c_0(\{P(T_g^\alpha(x))\}_{g \in K_0}) \neq Q(x)\})$

$$\leq \varepsilon/10 + \#K_0\varepsilon_0 + \varepsilon/5 + \varepsilon/10 < \varepsilon/2.$$

Since $m(\hat{\mu}) < \varepsilon_0 < \varepsilon/10$, this implies

$$\overline{m}(T_1^{\alpha_1}, Q_1; T^\alpha, Q) < \varepsilon,$$

and hence (T^α, Q) is m-f.d. □

Lemma 7.2.9. *Suppose (T^α, P) is m-f.d. and $Q : Z \to \Sigma_Q$ is $\mathscr{H}(T^\alpha, P)$-measurable. Then (T^α, Q) is m-f.d.*

Proof From Lemma 7.2.8, we know that $(T^\alpha, Q \vee P)$ is m-f.d. From Corollary 4.0.8 of the copying lemma, for any $\delta_0 > 0$ there exists δ so that if $(T_1^{\alpha_1}, Q_1)$ satisfies

(1) $\|(T_1^{\alpha_1}, Q_1), (T^\alpha, Q)\|_* < \delta$ and

(2) $h(T_1^{\alpha_1}, Q_1) > h(T^\alpha, Q) - \delta_0/6$,

then for $(Y, \mathscr{G}, \nu, S)$ a Bernoulli G-action of entropy at most $h(T^\alpha, P)$, there is a partition P_1 of $X_1 \times Y$ satisfying

(1) $\|(T^{\alpha_1} \times S, Q_1 \vee P_1), (T^{\alpha}, Q \vee P)\|_* < \delta_0$ and

(2) $h(T_1^{\alpha_1} \times S, Q_1 \vee P_1) > h(T^{\alpha}, Q \vee P) - \delta_0$.

This implies that

$$\overline{m}(T^{\alpha_1} \times S, Q_1 \vee P_1; T^{\alpha}, Q \vee P) < \varepsilon.$$

Thus restricting the m-joinings to the processes $(T_1^{\alpha_1}, Q_1)$ and (T^{α}, Q) we obtain

$$\overline{m}(T_1^{\alpha_1}, Q_1; T^{\alpha}, Q) < \varepsilon,$$

completing the proof that (T^{α}, Q) is m-f.d. □

Definition 7.2.10. *If $(X, \mathscr{F}, \mu, T^{\alpha})$ is m-f.d. for all finite partitions P we say T^{α} is m-f.d. Notice we now know this will be implied if (T^{α}, P) is m-f.d. for a generating partition P.*

We now show that m-f.d. is an m-equivalence invariant.

Theorem 7.2.11. *Suppose $(X, \mathscr{F}, \mu, T^{\beta})$ is m-f.d. Then for any α with $\alpha \overset{m}{\sim} \beta$, T^{α} is also m-f.d.*

Proof Fix a partition P and $\varepsilon > 0$. Suppose $\{\phi_i\} \subseteq \Gamma$ with $\alpha\phi_i \underset{m}{\rightarrow} \beta$. Set $\hat{\beta} = \langle\phi_i\rangle_{\alpha}$ in the m_{α}-closure of Γ. By Theorem 4.0.2 we can assume all the (α, ϕ_i) are bounded rearrangements. Choose ϕ_I so that

$$m_{\alpha}(\phi_I, \hat{\beta}) < \varepsilon/3.$$

Letting $\hat{\alpha} = \langle\phi_i^{-1}\rangle_{\beta}$ in the m_{β}-closure of Γ, we have

$$m_{\beta}(\phi_I^{-1}, \hat{\alpha}) = m_{\alpha}(\phi_I, \overline{\beta}) < \varepsilon/3.$$

As $T^{\beta\phi_I^{-1}}$ is conjugate to T^{β}, it is m-f.d. In particular $(T^{\beta\phi_I^{-1}}, P)$ is an m-f.d. process. Hence there exists $\delta > 0$ so that for any process $(X_1, \mathscr{F}_1, \mu_1, T_1^{\alpha_1}, P_1)$ satisfying

(1) $\|(T_1^{\alpha_1}, P_1), (T^{\beta\phi_I^{-1}}, P)\|_* < \delta$ and

(2) $h_m(T_1^{\alpha_1} P_1) > h_m(T^{\beta\phi_I^{-1}}, P) - \delta$ then

(3) $\overline{m}(T_1^{\alpha_1}, P_1; T_1^{\beta\phi_I^{-1}}, P) < \varepsilon_1$.

As $m_\alpha(\mathrm{id}, \beta\phi_I^{-1}) < \varepsilon/3$, there are $\psi_i \in FG(\mathrm{O})$ with $\alpha\psi_i \underset{m}{\rightarrow} \beta\phi_I^{-1}$ and

$$\sup_i m(\alpha, \psi_i) < \varepsilon/3.$$

By Lemma 5.0.7 we know

$$\liminf_{i\to\infty} h_m(T^{\alpha\psi_i}, P) \geq h_m(T^{\beta\phi_I^{-1}}, P),$$

(remember, $h_m = h$ as we are assuming the size is entropy-preserving) and certainly

$$\|(T^{\alpha\psi_i}, P), (T^{\beta\phi_I^{-1}}, P)\|_* \underset{i}{\rightarrow} 0.$$

Select ψ from among the ψ_i so that

(1) $\|(T^{\alpha\psi}, P), (T^{\beta\phi_I}, P)\|_* < \delta/2$ and

(2) $h_m(T^{\alpha\psi}, P) \geq h_m(T^{\beta\phi_I}, P) - \delta/2.$

Let $Q = P \vee P \circ \psi^{-1}$, and as $(T^{\alpha\psi}, P)$ and $(T^\alpha, P \circ \psi^{-1})$ are identical in distribution,

$$h(T^\alpha, Q) \geq h(T^{\alpha\psi}, P).$$

Choose a value δ_1 so that if

$$\|(\alpha_1, \psi'), (\alpha, \psi)\|_* < \delta_1$$

then we still will have

$$m(\alpha_1, \psi') < \varepsilon/3.$$

We proceed now as if m were an entropy-preserving size. If m is entropy-free, just replace the use of Corollary 4.0.11 with Corollary 4.0.9, without the need of the extra Bernoulli factor to obtain entropy.

By Corollary 4.0.11 there is a δ_0 so that if

(1) $\|(T_1^{\alpha_1}, P), (T^\alpha, P)\|_* < \delta_0$ and

(2) $h(T_1^{\alpha_1}, P) \geq h(T, P) - \delta_0$

then for $(Y, \mathscr{G}, \nu, S)$ a Bernoulli shift of sufficient entropy, there is a ψ' in the full-group of $U^{\alpha_1} = T_1^{\alpha_1} \times S$ so that:

(3) $\|(U^{\alpha_1\psi'}, P_1), (T^{\alpha\psi}, P)\|_* < \delta/2$

(4) $h(U^{\alpha_1\psi'}, P_1) \geq h(T^{\alpha\psi}, P) - \delta/2$ and

(5) $\|(\alpha_1, \psi'), (\alpha, \psi)\|_* < \delta_1/2.$

From this we can conclude that

$$\overline{m}(U^{\alpha_1\psi'}, P_1; T^{\beta\phi_I^{-1}}, P) < \varepsilon/3$$

and so

$$\overline{m}(T_1^{\alpha_1}, P_1; T^{\beta\psi_I^{-1}}, P) < 2\varepsilon/3$$

as $m(\alpha_1, \psi') < \varepsilon/3$.

Hence

$$\overline{m}(T_1^{\alpha_1}, P_1; T^{\alpha}, P) < \varepsilon$$

as $m_\alpha(\mathrm{id}, \overline{\beta}\psi_I^{-1}) < \varepsilon/3$, completing the proof that T^α is m-f.d. with respect to any partition P. □

7.3 The equivalence theorem

We now develop the background for and prove the equivalence theorem, that any two m-finitely determined G-actions of the same m-entropy are m-equivalent. Let $(X, \mathscr{F}, \mu, T^\alpha)$ and $(X_1, \mathscr{F}_1, \mu_1, T_1^{\alpha_1})$ be two free and ergodic G-actions. We already know that the space of measures $J_m(T_1^{\alpha_1}, T^\alpha)$ is a Polish topological space in the weak*-topology. What we intend to show is that if both of these actions are m-finitely determined and they have the same m-entropy then in fact those m-joinings that arise from m-equivalences between the two actions are a residual (dense G_δ) subset of $J_m(T_1^{\alpha_1}, T^\alpha)$.

Stated in this form, the equivalence theorem will follow directly from a corresponding "Sinai theorem" whose structure we now develop. Consider the subset $\mathscr{E}_m \subseteq J_m(T_1^{\alpha_1}, T^\alpha)$ defined as follows:

$$\mathscr{E}_m(T_1^{\alpha_1}, T^\alpha) = \{\hat{\mu} \in J_m(T_1^{\alpha_1}, T^\alpha) : \text{with respect to } \hat{\mu},$$

all f_i and hence ϕ_i are $p_1(\mathscr{F}_1)$-measurable and

$$p(\mathscr{F}) \subseteq p_1(\mathscr{F}_1)\}.$$

For $\hat{\mu}$ to belong to $\mathscr{E}_m$ means $\hat{\mu}$ can be thought of as follows. There are elements in the full-group of $T_1^{\alpha_1}$ for which

$$(f^{\alpha_1,\phi_1} \otimes f^{\alpha_1,\phi_2} \otimes \dots)^* \mu_1 \in \mathscr{M}^m.$$

Hence $\alpha_1\phi_i \underset{m}{\to} \beta$, and there is a measure-preserving map $\eta : X_1 \to X$ with $\eta T_1^\beta = T^\alpha \eta$. This really is, very loosely speaking, no more than saying α_1 is m equivalent to an arrangement β for which T_1^β has T^α as a factor. To actually obtain the joining in $\mathscr{E}_m$ one must select a sequence of full-group elements that achieves the m-equivalence, and then perhaps drop to a subsequence to guarantee they map μ_1 into $\mathscr{M}^m$. To see that any $\hat{\mu} \in \mathscr{E}_m$ gives rise to this situation, just notice that from the definition of $\mathscr{E}_m$, one can project the full-group elements ϕ_i from $X_1^G \times X^G \times \vec{\mathscr{R}}$ to X_1, as well as the identification of $\mathscr{F}$ as a subalgebra of $\mathscr{F}_1$.

Notice further that to have both $\hat{\mu} \in \mathscr{E}_m(T_1^{\alpha_1}, T^\alpha)$ and $\tilde{q}^*(\hat{\mu}) \in \mathscr{E}_m(T^\alpha, T_1^{\alpha_1})$ is to say that $\hat{\mu}$ arises from a pair of m-equivalent arrangements whose corresponding free and ergodic G-actions are conjugate to $T_1^{\alpha_1}$ and T^α respectively, i.e. the two actions are m-equivalent, and $\hat{\mu}$ is a detailed description of one such m-equivalence between them.

What we want to show first is that $\mathscr{M}_m(T_1^{\alpha_1}, T^\alpha)$ is a G_δ subset of $J_m(T_1^{\alpha_1}, T^\alpha)$ (perhaps empty). To complete the equivalence theorem what we will show is that if T^α is m-f.d., and $h_m(T_1^{\alpha_1}) \geq h_m(T^\alpha)$, then $\mathscr{E}_m(T_1^{\alpha_1}, T^\alpha)$ is a dense subset of $J_m(T_1^{\alpha_1}, T^\alpha)$.

To that end, for $I \in \mathbb{N}$, $P : X \to \Sigma_P$, a finite partition, and $\varepsilon > 0$ let

$$
\begin{aligned}
\mathscr{O}(I,P,\varepsilon) = \{\hat{\mu} \in J_m(T_1^{\alpha_1}, T^\alpha) :& \\
&\text{(i) there are } \phi_i' \quad i = 1,\dots,I \text{ in the full-group of } T_1^{\alpha_1} \text{ with} \\
&\quad \hat{\mu}(\{y = (p_1(x_1), p(x), \{f_i\}) : \phi_i'(x_1) \neq \phi_i(y)\}) < \varepsilon \text{ and} \\
&\text{(ii) there is a } P' : X_1 \to \Sigma_P \text{ with} \\
&\quad \hat{\mu}(\{(p_1(x_1), p(x), \{f_i\}) : P'(x_1) \neq P(x)\}) < \varepsilon\}.
\end{aligned}
$$

That is to say, an m-joining belongs to $\mathscr{O}(I,P,\varepsilon)$ if, loosely speaking, the first I full-group elements ϕ_i in the joining can be approximated by full-group elements measurable with respect to $\mathscr{F}_1$ (the first coordinate algebra) and the partition P, which is measurable with respect to $\mathscr{F}$ (the second coordinate σ-algebra), can be approximated by a partition in $\mathscr{F}_1$ (the first coordinate algebra).

Lemma 7.3.1. *The sets $\mathscr{O}(I,P,\varepsilon)$ are open subsets of $J_m(T_1^{\alpha_1}, T^\alpha)$.*

Proof Let $\hat{\mu} \in \mathscr{O}(I,P,\varepsilon)$. As the topological space $J_m(T_1^{\alpha_1}, T^\alpha)$ is a conjugacy invariant of the two G-actions, we can assume that the elements ϕ_i', $i = 1,\dots,I$ and the partition P' are all continuous functions of X_1 (in detail this is simply asking that a countable collection of sets be assumed clopen). Similarly we can assume that P is a continuous function of X. Now both (i) and (ii) of the definition of $\mathscr{O}(I,P,1/j)$ are simply asking that the $L^1(\hat{\mu})$ norms of a finite collection of continuous characteristic functions are $< \varepsilon$. Thus there will be a neighborhood of $\hat{\mu}$ in the weak*-topology on which both (i) and (ii) will continue to hold. □

Lemma 7.3.2. *Let P_i be a countable collection of finite Borel partitions of X, dense in the space of all Borel partitions taking values in $\mathbb{N}$. Then*

$$\mathscr{E}_m(T_1^{\alpha_1}, T^\alpha) = \cap_{i,I,j\in\mathbb{N}}\mathcal{O}(I, P_i, 1/j),$$

and hence is a G_δ.

Proof One containment follows directly from the definition, that

$$\mathscr{E}_m(T_1^{\alpha_1}, T^\alpha) \subseteq \cap_{i,I,j}\mathcal{O}(I, P_i, \mu).$$

To see the other containment, let $\hat{\mu} \in \cup_j\mathcal{O}(I, P_i, 1/j)$. There will be a partition P_i' of X and elements ϕ_i', $i = 1, \ldots, I$ in the full-group of $T_1^{\alpha_1}$ with

$$P_i' = P_i, \quad \hat{\mu}\text{-a.s. and}$$
$$\phi_i' = \phi_i, \quad i = 1, \ldots, I, \quad \hat{\mu}\text{-a.s.}$$

Hence for all $\hat{\mu} \in \cap_{i,I,j}\mathcal{O}(I, P_i, 1/j)$,

$$P_i' = P_i, \quad \hat{\mu}\text{-a.s. and}$$
$$\phi_i' = \phi_i, \quad \hat{\mu}\text{-a.s. for } \textbf{all} \text{ i.}$$

As the collection of sets in $p(\mathscr{F})$ that are also $\hat{\mu}$-a.s. also in $p_1(\mathscr{F}_1)$ forms a μ-complete σ-algebra which we have just shown contains all of the P_i, $p(\mathscr{F})$ is $\hat{\mu}$-a.s. a sub-σ-algebra of $p_1(\mathscr{F}_1)$. This proves the other containment

$$\cap_{i,I,j}\mathcal{O}(I, P_i, 1/j) \subseteq \mathscr{E}_m(T_1^{\alpha_1}, T^\alpha).$$

□

To complete the equivalence theorem we show that if $(X, \mathscr{F}, \mu, T^\alpha)$ is an m-f.d. free and ergodic G-action, then for any $(X_1, \mathscr{F}_1, \mu_1, T_1^{\alpha_1})$ with $h_m(T_1^{\alpha_1}) \geq h_m(T^\alpha)$, then each of the sets $\mathcal{O}(I, P, \varepsilon)$ is in fact dense in $J_m(T_1^{\alpha_1}, T^\alpha)$. Having already shown in Theorem 6.5.8 that $J_m(T_1^{\alpha_1}, T^\alpha)$ is a Polish space, we immediately conclude that $\mathscr{E}_m(T_1^{\alpha_1}, T^\alpha)$ is a dense subset of this space m-joinings. We refer to this proof of denseness as the "Sinai theorem" of our work as it embeds the m-f.d. action $T_1^{\alpha_1}$ as a factor of some T^β where $\alpha \overset{m}{\sim} \beta$.

Theorem 7.3.3. *Suppose $(X, \mathscr{F}, \mu, T^\alpha)$ is an m-f.d. free and ergodic G-action, and $(X_1, \mathscr{F}_1, \mu_1, T_1^{\alpha_1})$ is a free and ergodic G-action with*

$$h_m(T_1^{\alpha_1}) \geq h_m(T^\alpha).$$

Then for any $I \in \mathbb{N}$, finite partition P and $\varepsilon > 0$, the set $\mathcal{O}(I, P, \varepsilon)$ is dense in $J_m(T_1^{\alpha_1}, T^\alpha)$.

Proof Let $\hat{\mu} \in J_m(T_1^{\alpha_1}, T^\alpha)$ be some fixed m-joining, and let η be an open neighborhood of $\hat{\mu}$ in $J_m(T_1^{\alpha_1}, T^\alpha)$. Notice that in $J_m(T_1^{\alpha_1}, T_2^{\alpha_2})$ the arrangement $\boldsymbol{\alpha}$ represents α_1 and $\boldsymbol{\beta}$ represents α. We need to show that $\mathcal{O}(I, P, \varepsilon) \cap \eta$ is not empty.

As $J_m(T_1^{\alpha_1}, T^\alpha)$ is a conjugacy invariant, we can assume both X and X_1 are 0-dimensional spaces, that is to say have dense families of clopen sets, including the P-measurable sets. Hence we can find partitions Q of X_1 and P' of X, P' a refinement of P, a value $I_0 \in \mathbb{N}$ and a value ε_0, $\varepsilon > \varepsilon_0 > 0$ so that if $\hat{\mu}_1 \in J_m(T_1^{\alpha_1}, T^\alpha)$ satisfies

$$\text{(a)} \quad \|\Big(\sigma^{\boldsymbol{\alpha}}, Q \circ \pi_1 \vee P' \circ \pi_2 \vee \bigvee_{i=1}^{I_0} g_{(\boldsymbol{\alpha}, \phi_i)}\Big)_{\hat{\mu}},$$

$$\Big(\sigma^{\boldsymbol{\alpha}}, Q \circ \pi_1 \vee \bigvee_{i=1}^{I_0} g_{(\boldsymbol{\alpha}, \phi_i)}\Big)_{\hat{\mu}_1} \|_* < \varepsilon_0$$

then $\hat{\mu}_1 \in \eta$. We will construct $\hat{\mu}_1$ satisfying this as well as

$$\text{(b)} \quad \hat{\mu}_1 \in \mathcal{O}(I_0, P', \varepsilon_0) \subseteq \mathcal{O}(I, P, \varepsilon).$$

To obtain $\hat{\mu}_1$ we will construct an I_0, J, δ-perturbation of the free and ergodic G-action

$$((X_1 \times X)^G \times \vec{\mathscr{R}}, \mathscr{B}, \hat{\mu}, \sigma^{\boldsymbol{\alpha}}).$$

This perturbation will be the m-joining we want in the loose sense. We then just map it by Q^* into the space of m-joinings.

By Theorem 7.1.5 there exists $\delta_1 > 0$ such that for any J_1 sufficiently large, any I_0, J_1, δ_1-perturbation of the above system will still project to a measure in $\mathscr{M}^m$. Our choices for δ and J will be at least this small and large respectively, and hence all we need see is how to also obtain both (a) and (b) above via a perturbation.

As (T^α, P') is m-f.d., there is a $\delta_2 > 0$ so that if $T_2^{\alpha_2}$ is some free and ergodic G-action with:

(1.0) $\|(T_2^{\alpha_2}, P_2'), (T^\alpha, P')\|_* < \delta_2$ and

(2.0) $h_m(T_2^{\alpha_2}) > h_m(T^\alpha, P') - \delta_2$ then

(3.0) $\overline{m}(T_2^{\alpha_2}, P_2'; T^\alpha, P') < \delta_1$.

As $\boldsymbol{\alpha\phi}_i \to \boldsymbol{\beta}$ in the m-joining $\hat{\mu}$ and $(\sigma^{\boldsymbol{\beta}}, P')_{\hat{\mu}}$ is identical to (T^α, P') in distribution, we must have both

(1.1) $\|(\sigma^{\boldsymbol{\alpha\phi}_j}, P'), (T^\alpha, P')\|_* \underset{j}{\to} 0$ and

(2.1) $h_{m,\hat{\mu}}(\sigma^{\boldsymbol{\alpha\phi}_j}, P') \underset{j}{\to} h_m(T^\alpha, P')$.

Select J_1 larger if necessary to ensure:

(1.2) $\|(\sigma^{\boldsymbol{\alpha\phi}_{J_1}}, P'), (T^\alpha, P')\|_* < \delta_2/2$ and

(2.2) $h_{m,\hat{\mu}}(\sigma^{\boldsymbol{\alpha\phi}_{J_1}}, P') > h_m(T^\alpha, P') - \delta_2/2$.

As $h_m(T_1^{\alpha_1}) \geq h_m(T^\alpha)$, select a finite partition Q'' of X_1 refining Q' with

$$h_m(T_1^{\alpha_1}, Q'') > h_m(T^\alpha) - \delta_2/3.$$

Consider the free and ergodic G-action $\sigma^{\boldsymbol{\alpha}}$ relative to the measure $\hat{\mu}$ and partitions Q'', P' and the finite list of full-group elements $\phi_1, \phi_2, \ldots, \phi_{I_0}, \phi_{J_1}$. We are precisely in the position to apply Theorem 4.0.12 to copy in all of these full-group elements along with the partition P'. Construct these copies sufficiently close in distribution and in m-entropy (which here is of course just entropy) to obtain all of the following.

(i) $D((f^{\alpha,\phi_1} \otimes \cdots \otimes f^{\alpha,\phi_{I_0}} \otimes f^{\alpha,\phi_{J_1}})^* \hat{\mu},$

$$(f^{\alpha_1,\phi_1'} \otimes \cdots \otimes f^{\alpha_1,\phi_{I_0}'} \otimes f^{\alpha_1,\phi_{I_0+1}'})^* \mu_1) < \delta_1$$

(1.3) $\|(T^{\alpha_1\phi_{I_0+1}'}, P_1), (T^\alpha, P')\|_* < \delta_2$ from (1″).

(2.3) $h_m(T_1^{\alpha_1\phi_{I_0+1}'}, P_1) > h_m(T^\alpha, P') - \delta_2$

and hence we will have

(3.3) $\overline{m}(T_1^{\alpha_1\phi_{I_0+1}'}, P_1; T^\alpha, P') < \delta_1$.

We also require of the copying that

(a)′ $\|(\sigma^{\boldsymbol{\alpha}}, Q'' \circ \pi_1 \vee P' \circ \pi_2 \vee \bigvee_{i=1}^{I_0} g_{(\boldsymbol{\alpha},\phi_i)})_{\hat{\mu}},$

$$(T_1^{\alpha_1}, Q'' \vee P_1' \vee \bigvee_{i=1}^{I_0+1} g_{(\alpha_1,\phi_i')})\|_* < \varepsilon_0.$$

Now (3.3) tells us there is an m-joining $\hat{\mu}_0$ of $T_1^{\alpha_1\phi'_{I_0+1}}$ and T^α with

(1) $\hat{\mu}_0(P_1 \triangle P') < \delta_1 < \varepsilon_0$ and

(2) $m(\hat{\mu}_0) < \delta_1$.

We can lift to this m-joining of $T_1^{\alpha_1\phi'_{I_0+1}}$ and T^α the full-group elements ϕ'_i, and the arrangement α_1 yielding the sequence of G-actions

$$\sigma^{\alpha_1}, \sigma^{\alpha_1\phi'_1}, \ldots, \sigma^{\alpha_1\phi'_{I_0+1}} \text{ where } \boldsymbol{\alpha} = \alpha_1\phi'_{I_0+1}.$$

The m-joining now comes equipped with its canonical full-group elements $\boldsymbol{\phi}_i$ with $\boldsymbol{\alpha\phi}_i \underset{i}{\to} \boldsymbol{\beta}$ and for which

$$\limsup_{i\to\infty} m_{\hat{\mu}_0}(\boldsymbol{\alpha}, \boldsymbol{\phi}_i) < \delta_1$$

which is to say

(ii) $m_{\alpha_1\phi_{I_0+1},\hat{\mu}_0}(\langle\phi_i\rangle, \mathrm{id}) < \delta_1$.

Notice that (i) and (ii) tell us that the G-action $((X_1 \times X)^G \times \vec{\mathscr{R}}, \mathscr{B}, \hat{\mu}_0, \sigma^{\alpha_1})$ with full-group elements $\phi'_1, \ldots, \phi'_{I_0+1}$ and $\{\boldsymbol{\phi}_i\}$ form an I_0, J_1, δ_1-perturbation of the original m-joining $\hat{\mu}_1$ with its canonical list of full-group elements. In particular we know by Lemma 7.1.5 that dropping to a subsequence of the $\boldsymbol{\phi}_i$ if necessary we will have

$$(f^{\alpha_1,\phi''_1} \otimes f^{\alpha_1,\phi''_2} \otimes \cdots \otimes)^* \hat{\mu}_0 \in \mathscr{M}^m.$$

Examining this perturbation more closely, consider the two projections:

$$\pi_1(\{\vec{x}\}, \{\vec{x}_1\}, \{f_1\}) \to \vec{x}(\mathrm{id}) \text{ and}$$
$$\pi_2(\{\vec{x}\}, \{\vec{x}_1\}, \{f_1\}) \to \vec{x}_1(\mathrm{id}).$$

These satisfy:

$$\pi_1^*(\hat{\mu}_0) = \mu_1, \quad \pi_2^*(\hat{\mu}_0) = \mu,$$
$$\pi_1\sigma^{\alpha_1} = T_1^{\alpha_1}\pi_1 \text{ and } \pi_2\sigma^\beta = T^\alpha\pi_2$$

and hence this is an m-joining of $T_1^{\alpha_1}$ and T^α in the loose sense. Let

$$\hat{\mu}_1 = Q^*(\hat{\mu}_0)$$

be the corresponding m-joining in $J_m(T_1^{\alpha_1}, T^\alpha)$.

Notice that $\phi''_i = \phi'_i$ for $i = 1, \ldots, I_0$ are all $\pi_1^{-1}\mathscr{F}_1$-measurable, and as $\hat{\mu}_0(P_1 \triangle P') < \varepsilon_0$, we will have (b), that

$$\hat{\mu}_1 \in \mathscr{O}(I_0, P', \varepsilon_0).$$

Statement (a′) implies directly that $\hat{\mu}_1$ satisfies (a) and the result is proven. □

Appendix

Our intent in this appendix is to provide a linkage to the two previous papers the authors have written on restricted orbit equivalence. We say that an r-size is a size function m as defined in this current work, in Section 2.2. (The "r" denotes "rearrangement".) We will see that the notion of m-equivalence developed in [25] (where m is a p-size, where "p" denotes "permutation") is subsumed by our work here. On the other hand, we will not quite be able to show this for the work in [43]. What we will see though is that a slight strengthening in the definition of the equivalence relation associated with a 1-size, (size as defined in [43]) will make it possible to describe the equivalence relation as a restricted orbit equivalence in the sense we describe here. As we will see, this change will have no effect on the examples described in [43], and the m-f.d. systems for the original equivalences are unchanged by this strengthening. Whether some equivalence classes of arrangements are possibly changed for some 1-size, we do not know. As we now consider the definition in [43] to have been a very preliminary, perhaps unrefined, attempt to axiomatize the notion of restricted orbit equivalence, we have not pursued this issue further. Our main interest here is to bring the examples, in particular examples 3 and 4 (referred to as m_ψ and m_ϕ) under the umbrella of our work here.

We break this work into two sections. First we will handle the notion of a size, here called a 1-size (for 1-dimensional) used in [43] for actions of $\mathbb{Z}$. This will be our most detailed section. Then we will discuss the notion of size first put forward in [25], here called p-sizes, for actions of $\mathbb{Z}^d$. In so doing we will also consider a preliminary axiomatization we have given for discrete amenable group actions, as it lies somewhere between that of [25] and what we have discussed here.

What we do here is rather technical, and as it is perhaps not of great interest to the general reader, we will not provide a great deal of motivation. We also will assume that an interested reader has a copies of [25] and [43] in hand to refer to.

A.1 1-sizes

The notion of a 1-size, as set up in [43] begins at the level of permutations π of intervals of integers $(i, i+1, \dots, j)$, and lifts from here to bijections of $\mathbb{Z}$ via the definition

$$m(f) = \liminf_{\substack{i \to -\infty \\ j \to \infty}} m(\pi_{f,(i,j)})$$

where $\pi_{f,(i,j)}$ is the "push-together" of $f|_{(i,j)}$, i.e. the permutation of (i, j) that reorders points exactly as $f|_{(i,j)}$ does. In [43] a bijection of $\mathbb{Z}$ is indicated by a function f. To any such bijection f there corresponds a unique bijection fixing 0 that reorders points exactly as f does:

$$h(n) = f(n) - f(0)$$

(as we are in $\mathbb{Z}$ we use additive notation). Axiom (ii) (page 7 of [43]) requires that m be "stationary", i.e. as a calculation on a permutation it should depend only on the number of terms and the way they are reordered, not on where the block (i, j) is placed. Thus the calculation $m(f)$ is translation invariant (regarding f as an element of $\mathbb{Z}^{\mathbb{Z}}$), $m(h) = m(f)$. This then implies, using our notation, that

$$m(h) = m(S^{\alpha}(h)) \text{ as } S^{\alpha}(h)(n) = \sigma(f)(n) - \sigma(f)(0).$$

The value of the 1-size distance between a pair of arrangements $(m(\alpha_1, \alpha_2))$ (called orderings on $\mathbb{Z}$) is now set to be the a.s. constant value of $m(h_x^{\alpha_1,\alpha_2})$.

Notice that we could have written this as

$$m(\alpha_1, \alpha_2) = \int m(h_x^{\alpha_1,\alpha_2})\, d\mu.$$

Written this way, we see this can be viewed as the integral with respect to various invariant measures of the Borel function $m(h)$ on $\mathcal{G}$.

Two arrangements are defined to be m-equivalent (written $\alpha_1 \overset{m}{\sim} \alpha_2$) in [43] if there are full-group elements ϕ_i and $m(\alpha_1\phi_i, \alpha_2) \underset{i}{\to} 0$. The given m is not a metric (soon we will see how easily it can be made one) but taking it to be one, what we are doing is taking the m-closure of the set of all $\alpha_1\phi$ in the full-group.

This approach is most different from our current one in that it evaluates a distance between arbitrary arrangements rather than just rearrangement pairs (α, ϕ). This is best understood by the introduction of the *collapsing* of one arrangement onto another, which converts a pair of arrangements into a rearrangment. This has a serious fault, that the collapsing of a pair $\alpha, \alpha\phi$ will produce a new pair $\alpha, \alpha\phi'$ where ϕ and ϕ' are not necessarily close in L^1. This leads to a serious difficulty in obtaining Axiom 3 for the r-size we construct. This is more than just technical, as Axiom 3 plays a pivotal role in all of our development. It did as well in the development of [43], in particular in trying to develop the *m*-f.d. notion. This is the motivation for the consideration there of *dividing* rearrangements, and the restrictive definition of an *m*-joining, requiring the two processes to be covered by a third in which they sit linked by a *bounded rearrangement.* Lying behind this is the fact that if (α, ϕ) is in fact bounded then collapsings (α, ϕ') can be chosen such that ϕ and ϕ' are L^1-close (Corollary 4.5 of [43]).

This problem that a collapsing of a rearrangement is not necessarily a small L^1-change is also hidden in the interplay between the definition of $m(\alpha_1, \alpha_2)$ via collapsing of blocks and the requirement that m satisfy an approximate triangle inequality. For these to coexist for the same notion of m forces severe restrictions, and is in some sense the reason we can show all of our examples are equivalent to r-sizes, and also the reason that α-equivalence [8] comes from a p-size but not a 1-size.

We introduce two ideas to provide the linkage we wish to demonstrate between 1-sizes and r-sizes.

Definition A.1.1. *We say a rearrangement (α, ϕ) is bounded if the function $f_x^{\alpha,\phi} = \alpha(x, \phi(x))$ is bounded μ-a.s. We say a rearrangement (α, ϕ) is blocked and bounded (abbreviated B&B) over a subset F if the return-time $r_F(x) = \min(n > 0 : T_n^\alpha(x) \in F)$ is bounded, and ϕ acts as a permutation of each return-time block*

$$(x, T_1^\alpha(x), \dots, T_{r_F(x)-1}^\alpha(x)), x \in F.$$

Lemma A.1.2. *Those ϕ with (α, ϕ) bounded form a subgroup of the full-group and Theorem 4.0.2 tells us that those ϕ with (α, ϕ) B&B over some set are $\|\cdot,\cdot\|_s^\alpha$-dense in the full-group.*

Definition A.1.3. *Axiom (v) of a 1-size tells us that for any $\varepsilon > 0$ there are δ and N so that if (α, ϕ) is B& B over some set F with*

(1) $r_F(x) \geq N$ *for all* $x \in F$ *and*
(2) $m(\pi_{x,(0,r_F(x)-1)}) < \delta$ *then*

$m(\alpha, \alpha\phi) < \varepsilon$. *Any* (α, ϕ) *B&B over a set* F *satisfying* (1) *and* (2) *we will call* m, ε*-B&B.*

Definition A.1.4. *We say a sequence of rearrangements* (α, ϕ_i) *is nicely-blocked over* F_1 *if each of the rearrangements* $(\alpha\phi_i, \phi_i^{-1}\phi_{i+1})$ *is B&B over a set* F_i *where* $F_{i+1} \subseteq F_i$. *Moreover on each return time block*

$$(x, T_1^\alpha(x), \ldots, T^\alpha_{r_{F_i}(x)-1}(x)), x \in F_i$$

the next full-group element ϕ_i *acts as a fixed power of* T^α. *That is to say, for each* $x \in F_i$ *there is a value* $j(x)$ *and for all* x' *in the return-time block over* x, $\phi_{i+1}(x') = T^\alpha_{j(x)}(x')$.

We say the sequence of rearrangements is m-nicely-blocked over F_i *if for all* $i \leq j$, $(\alpha\phi_i, \phi_i^{-1}\phi_j)$ *is B&B over* F_i *and is in fact* $m, 2^{-i} - 2^{-j}$*-B&B.*

Lemma A.1.5. *If* (α, ϕ_i) *is nicely-blocked over the sequence of sets* F_i *and* $\mu(F_i) \underset{i}{\to} 0$ *then* $\alpha\phi_i$ *is converging in* L^1 *to an arrangement* β. *Furthermore the sequence* (β, ϕ_i^{-1}) *is also nicely-blocked over the sequence of sets* $\phi_i(F_i)$ *and* $\beta\phi_i^{-1} \underset{i}{\to} \alpha$ *in* L^1. *We get no symmetry like this for m-nicely-blocked sequences as m is only assumed symmetric for arrangements and not for permutations in that knowing* $m(\pi)$ *is small does not a-priori imply that* $m(\pi^{-1})$ *is small.*

Definition A.1.6. *For* m *a* 1*-size we say* $\alpha_1 \overset{m}{\sim} \alpha_2$ *nicely if there is an m-nicely-blocked sequence of rearrangements* ϕ_i *with*

$$m(\alpha_1\phi_i, \alpha_2) \underset{i}{\to} 0.$$

This relation cannot be assumed to be either symmetric or transitive.

Definition A.1.7. *We say a* 1*-size* m *is a* 1^+*-size if for any* $\varepsilon > 0$ *there is a* δ *and for any rearrangement* (α, ϕ) *with* $m(\alpha, \alpha\phi) < \delta$, *and any* $\varepsilon_1 > 0$ *there is a bounded* ϕ' *with*

1. $\mu(\{x : \phi(x) \neq \phi'(x)\}) < \varepsilon_1$ *and*
2. $m(\alpha, \alpha\phi') < \varepsilon$.

Theorem A.1.8. *All of the examples of* 1*-sizes discussed in* [43] *are* 1^+*-sizes.*

Proof Note: In these arguments we will assume the reader has access to [43]. The two examples (m_0 and m_∞) and two classes of examples (m_ψ and m_ϕ) all have the common feature that $m(\alpha_1, \alpha_2)$ can be defined without reference to collapsing. For m_0 this is done in Lemma 2.6 of [43], for m_ψ in Lemma 2.10 of [43], for m_ϕ in Lemma 2.14 of [43], and notice for m_∞, if $m_\infty(\alpha_1, \alpha_2) < 1$ then $\alpha_1 = \alpha_2$. Modifications of their proofs give the arguments we seek.

We really only need consider m_ψ and m_ϕ. Suppose we consider $\alpha = \alpha_1$ and $\alpha\hat{\phi} = \alpha_2$. Theorem 4.0.2 gives a method to modify $\hat{\phi}$ by less than any preassigned amount in L^1 to a $\hat{\phi}'$ so that $(\alpha, \hat{\phi}')$ is bounded. We show that given any $\varepsilon > 0$ there is a δ so that if the original $m(\alpha, \alpha\hat{\phi}) < \delta$ then the new value $m(\alpha, \alpha\hat{\phi}') < \varepsilon$. When we refer to a *tower block* or block of the tower here we mean a consecutive sequence of orbit points that begins at the base of the tower and moves up to the top of the tower.

For m_ψ, consider the construction in Theorem 4.0.2 in which the rearrangement $(\alpha, \hat{\phi})$ is modified to a bounded rearrangement $(\alpha, \hat{\phi}')$ by cutting the orbit into tower blocks by a Rokhlin tower, and taking those points thrown out of a block by $\hat{\phi}$ and mapping them to those points whose preimages are outside the block. We describe now why $m(\alpha, \alpha\hat{\phi}')$ must be small if $m(\alpha, \alpha\hat{\phi})$ is. In fact one can use the same set A of Lemma 2.10 of [43] to show this. Consider an interval (u, v) as a block of points in the orbit of some point x. This segment of orbit will be covered by a sequence of Rokhlin tower blocks, with two perhaps partial blocks at the end, which we write as (u, u_1) and (v_1, v). The cardinality of the set $f_x^{\alpha,\alpha\hat{\phi}'}((u, v))\triangle(u, v)$ is twice the cardinality of the set $B'(u, v)$ consisting of those points in (u, v) whose image under $f_x^{\alpha,\hat{\phi}'}$ is not in (u, v). Now the only points that can possibly be in $B'(u, v)$ are those in (u, u_1) or (v_1, v). Let $B(u, v)$ be those points in (u, v) that are thrown out by the original map $f_x^{\alpha,\hat{\phi}}$. Notice that if $j \in B'(u, v)$ but $j \notin B(u, v)$ it must be the case that j originally was thrown out of the Rokhlin tower block containing $T_j^\alpha(x)$, but to a point closer to x on the orbit, and when $\hat{\phi}'$ was constructed, it was moved to a point further out away from x but inside the Rohklin tower block containing $T_j^\alpha(x)$. This means j lies outside (u_1, v_1) but its image under $f_x^{\alpha,\hat{\phi}}$ lies inside (u_1, v_1). This implies that one of the two calculations

$$\#(f_x^{\alpha,\hat{\phi}}((u, v))\triangle(u, v)) \quad \text{or} \quad \#(f_x^{\alpha,\hat{\phi}}((u_1, v_1))\triangle(u_1, v_1))$$

is at least half the size of

$$\#(f^{\alpha,\hat{\phi}'}((u, v))\triangle(u, v)).$$

As $\psi(n)/\psi(2n)$ is required to be bounded away from zero in n, we obtain the result.

For m_ϕ, work from the characterization of m_ϕ in Lemma 2.14 of [43]. Note that we will represent full-group elements here by $\hat{\phi}$ and $\hat{\phi}'$ to avoid misunderstanding with the use of ϕ as a parameter of m. To begin, fix $1/4 > \varepsilon > 0$ and select a value B_0 so that

$$\mu(\{x : |\alpha(x, \hat{\phi}(x))| < B_0\}) > 1 - \varepsilon/10.$$

Next choose a value B_1 so that for any $|i| \geq B_1$,

$$\frac{\phi(i + B_0)}{\phi(i)} < 1 - \varepsilon/10 \quad \text{and} \quad \frac{\phi(\frac{\varepsilon}{10} i)}{\phi(i)} > 1 - \varepsilon/10.$$

Next choose a value B_2 so that

$$\mu(\{x : |\alpha(x, \hat{\phi}(x))| < B_2\}) > 1 - \frac{\varepsilon}{20B_1}.$$

Thus:

$$\mu(\{x : f_x^{\alpha,\phi}((-B_1, B_1)) \subseteq (-B_2 - B_1, B_2 + B_1)\}) > 1 - \varepsilon/10.$$

Choose N so large that $N\varepsilon/20 > B_2 + B_1$.

In Theorem 4.0.2 be sure the Rokhlin tower has height $H \geq N$ and covers all but $\varepsilon/10$ in measure of X. Apply the construction of Theorem 4.0.2 to this tower. That is, map those points x in the tower whose image under $\hat{\phi}$ throws the point out of the tower block containing x to those points in this tower block that do not have $\hat{\phi}$-preimages in it. On points outside the tower, we replace $\hat{\phi}$ with the identity, to create the new full-group element $\hat{\phi}'$.

It is a direct calculation from our choices for B_0, B_1, B_2, N and H that $\mu(\{x : \hat{\phi}(x) \neq \phi(x)\}) < \varepsilon$. To see that $m_\phi(\alpha, \alpha\hat{\phi}')$ is still small, we assume $m_\phi(\alpha, \alpha\hat{\phi}) < \delta < 1/4$ and let A be the set given by Lemma 2.14 of [43].

Remove from A all points that are

(1) outside the tower,
(2) within $\varepsilon H/5$ of either end of a tower block, or for which
(3) $h_x^{\alpha,\alpha\hat{\phi}}|_{(-B_1,B_1)} \neq h_x^{\alpha,\alpha\hat{\phi}'}|_{(-B_1,B_1)}$.

Call the remaining set A'. It follows from our estimates that

$$\mu(A') > \mu(A) - \varepsilon.$$

We wish to show that this set now can be used in Lemma 2.14 of [43] to show that $m_\phi(\alpha, \alpha\hat{\phi}') < 3\delta + \varepsilon$. We do this by showing that for any $x \in A'$ and any x' in the orbit of x, that

$$\left|\frac{\phi(|\alpha(x, x')|)}{\phi(|\alpha(\hat{\phi}'(x), \hat{\phi}'(x'))|)} - 1\right| < \delta + \varepsilon/3.$$

Note: if $|a - 1| < d < 1/2$ then $|a - \frac{1}{a}| < 3d$ and if $|a - \frac{1}{a}| < d$ then $|a - 1| < d$.

We only need consider those x' with $\hat{\phi}(x') \neq \hat{\phi}'(x')$ as otherwise we already have the estimate. Hence we can assume $|\alpha(x, x')| > B_1$. Suppose x' is such a point. There are two possibilities:

(1) $\hat{\phi}'(x') = x'$ or
(2) x' lies in a tower block, but is thrown out of it by $\hat{\phi}$, and hence $x'' = \hat{\phi}(x')$ lies in this tower block, but $\hat{\phi}^{-1}(x'')$ does not.

To understand (1) notice

$$\alpha(\hat{\phi}'(x), \hat{\phi}'(x')) = \alpha(x, x') + \alpha(\hat{\phi}'(x), x).$$

As $|\alpha(\hat{\phi}'(x), x)| \leq B_0$ and $|\alpha(x, x')| > B_1$, we conclude:

$$\left|\frac{\phi(|\alpha(\hat{\phi}'(x), \hat{\phi}'(x'))|)}{\phi(|\alpha(x, x')|)} - 1\right| < \varepsilon/10.$$

To understand (2) we describe the case when all three points x, x' and x'' lie in the same tower block. When they do not, the estimates improve. Once more there are three possibilities:

(1) Both $|\alpha(x, x')|$ and $|\alpha(\hat{\phi}(x), x'')| > \varepsilon H/10$, or

(2) $|\alpha(x, x')| \leq$ both $|\alpha(\hat{\phi}(x), x'')|$ and $\varepsilon H/10$ or

(3) $|\alpha(\hat{\phi}(x), x'')| \leq$ both $|\alpha(x, x'))|$ and $\varepsilon H/10$.

In case (1) we will have that $\frac{|\alpha(x,x')|}{|\alpha(\hat{\phi}(x),x'')|}$ must lie between $\varepsilon/10$ and $10/\varepsilon$, both numerator and denominator are larger than B_1 and we conclude that

$$\left|\frac{\phi(|\alpha(x, x')|)}{\phi(|\alpha(\hat{\phi}'(x), \hat{\phi}'(x')|)} - 1\right| < \varepsilon/10.$$

Cases (2) and (3) will be handled in a parallel fashion. For (2), we will have $|\alpha(\hat{\phi}(x), \hat{\phi}(x'))| > \varepsilon H/10$ but $|\alpha(\hat{\phi}(x), x'')| \leq H$ and so:

$$1 \geq \frac{|\alpha(x, x')|}{|\alpha(\hat{\phi}(x), x'')|} \geq \frac{\varepsilon}{10} \frac{|\alpha(x, x')|}{|\alpha(\hat{\phi}(x), \hat{\phi}(x'))|}.$$

Thus

$$\frac{\phi(|\alpha(x,x')|)}{\phi(\frac{10}{\varepsilon}|\alpha(\hat{\phi}(x),\hat{\phi}(x'))|)} = \frac{\phi(|\alpha(x,x')|)}{\phi(|\alpha(\hat{\phi}(x),\hat{\phi}(x'))|)}\frac{\phi(|\alpha(\hat{\phi}(x),\hat{\phi}(x'))|)}{\phi(\frac{10}{\varepsilon}|\alpha(\hat{\phi}(x),\hat{\phi}(x'))|)}$$
$$> (1-\delta)(1-\varepsilon/10) > 1-\delta-\varepsilon/10.$$

For (3) use the same argument as for (2) but from the point of view of $\hat{\phi}^{-1}$, i.e. replace x with $\hat{\phi}(x)$ and x' with x'' and $\hat{\phi}$ with $\hat{\phi}^{-1}$ throughout. This completes the calculation and the proof. □

Having seen that the examples of [43] are 1^+-sizes, we now want to explain the link between nice m-equivalences and 1^+-sizes. To begin we now remind the reader in more detail of the construction of collapsings. This is carried out on pages 47–48 of [43]. By a *collapsing* $\overline{\alpha}_2$ of α_2 on α_1 one means the construction of a full-group element ϕ ($\overline{\alpha}_2 = \alpha_1\phi$) so that (α_1, ϕ) is B&B over a set F and for each $x \in F$,

$$\pi_{f_x^{\alpha_1,\alpha_2},(0,r_F(x)-1)} = \begin{cases} \pi_{f_x^{\alpha_1,\overline{\alpha}_2},(0,r_F(x)-1)} & \text{or} \\ \text{id} & \end{cases}.$$

That is to say, $\overline{\alpha}_2$ is obtained by *pushing together* or *collapsing* the image points $f_x^{\alpha_1,\alpha_2}((0,\ldots,r_F(x)-1))$ into a consecutive block. It is convenient to allow the collapsing to act as the identity on some of these blocks. Such a collapsing is called an $\hat{\varepsilon}, K$-collapsing if (in our vocabulary)

$$\mu(\{x : h_x^{\alpha_1,\alpha_2}|_{(-K,K)} = h_x^{\alpha_1,\overline{\alpha}_2}|_{(-K,K)}\}) > 1-\hat{\varepsilon}.$$

Note: Having $\hat{\varepsilon}$ in the definition in [43] will give an $\hat{\varepsilon}$ in this definition. Having an $\hat{\varepsilon}$ in this definition will give $\sqrt{\hat{\varepsilon}}$ in the [43] definition.

Definition A.1.9. *In* [43] *the notion of an* $\varepsilon, \hat{\varepsilon}, K$*-collapsing was also given where one also asked that*

$$m(\alpha_1, \overline{\alpha}_2) < \varepsilon.$$

We give here a stronger definition for this notion than the one actually used in [43]. *Notice that an* $\hat{\varepsilon}, K$*-collapsing is a B&B rearrangement. If it is in fact an* m, ε*-B&B arrangement we call it an* $\varepsilon, \hat{\varepsilon}, K$*-collapsing (see Definition A.1.3). This in particular forces* $m(\alpha_1, \alpha_1\phi) < \varepsilon$.

Lemma A.1.10 (Lemma 3.2 of [43]**).** *Given any* $\varepsilon > 0$ *there is a* δ *so that if* $m(\alpha_1,\alpha_2) < \delta$ *then for all* $\hat{\varepsilon}, K$ *there exists an* $\varepsilon, \hat{\varepsilon}, K$*-collapsing* $\overline{\alpha}_2$ *of* α_2 *on* α_1.

Proof Although [43] only claims to obtain its definition of an $\varepsilon, \hat{\varepsilon}, K$-collapsing, it does so by obtaining a collapsing satisfying our Definition A.1.9. □

Collapsings behave particularly nicely with respect to bounded rearrangements (or more generally what are called dividing reorderings in [43]). We recall a few simple facts about rearrangements of $\mathbb{Z}$. Recall that $\alpha\phi = \alpha\phi'$ iff $\phi' = T_j^\alpha \circ \phi$ (we are assuming ergodicity universally). Next recall that for (α, ϕ) a bounded rearrangement,

$$\int \alpha(x, \phi(x))\, d\mu$$

is always an integer, which we call $J(\alpha, \phi)$ (the average translation induced by ϕ). Thus we can replace ϕ by $\phi' = T^\alpha_{-J(\alpha,\phi)}\phi$ to obtain a rearrangement (α, ϕ') with $J(\alpha, \phi') = 0$ and $\alpha\phi = \alpha\phi'$. As the calculation of m for a rearrangment is $m(\alpha, \alpha\phi)$, this change is invisible to m, and hence we can always assume our rearrangements are such that $J(\alpha, \phi) = 0$. Also notice that for a collapsing $\alpha_1\phi$ of any α_2 on α_1 we always have $J(\alpha, \phi) = 0$. The next lemma explains the value of this.

Lemma A.1.11. *If (α, ϕ) is a bounded rearrangement with $J(\alpha, \phi) = 0$ then for the collapsing $\alpha\phi'$ of $\alpha\phi$ onto α over a set F,*

$$\mu(\{x : \phi(x) \neq \phi'(x)\}) \leq 2\mu(F)\|\alpha(x, \phi(x))\|_\infty.$$

Proof See Lemma 4.3 and Corollaries 4.4 and 4.5 of [43]. □

We now need a little trick which we will discuss in more detail later under the notion of a *convergence criterion.* For now we need just a very special case.

Lemma A.1.12. *Suppose $(\alpha, \phi_i), i = 1, \ldots, N$ is a finite sequence of blocked and bounded rearrangements which for all $i \leq j$, $(\alpha\phi_i, \phi_i^{-1}\phi_j)$ is $m, 2^{-i}-2^{-j}$-B&B. Assume further that they are nice in that the base sets F_i over which they are blocked are nested and on any return-time block of F_i the rearrangement $(\alpha\phi_i, \phi_i{}^{-1}\phi_{i+1})$ acts by a constant translation. (This is saying that the finite list is the beginning of a potentially m-nicely-blocked sequence.)*

There is then a δ_N depending on this finite list so that for any bounded rearrangement (α, ϕ_{N+1}) with $J(\alpha, \phi_{N+1}) = 0$ and

$$m(\alpha\phi_N, \alpha\phi_{N+1}) < \delta$$

we will be able to construct a new list (α, ϕ_i'), $i = 1, \ldots, N+1$ *of blocked and bounded rearrangements, satisfying all the conditions of the original list but with N replaced by* $N+1$ *and with, for all* $i = 1, \ldots, N+1$,

$$\mu(\{x : \phi_i(x) \neq \phi_i'(x)\}) \leq 2^{-N-9}.$$

Proof Notice that if δ is small enough, we can obtain $\alpha\phi'_{N+1}$ as a collapsing of $\alpha\phi_{N+1}$ on $\alpha\phi_N$ with all its conditions, and its base set F_{N+1} as small as we like. Set $\psi = \phi_{N+1}^{-1}\phi'_{N+1}$ and define $\phi_i'' = \phi_i\psi$ for all i. We can assume $\mu(\{x : \psi(x) \neq x\}) \leq 2^{-N-10}$. The rearrangements ϕ_i'' are B&B with base sets $F_i'' = \psi^{-1}(F_i)$. The base set F_{N+1} may not lie in the various F_i'', $i \leq N$. This is really all we lack. For each $x \in F_{N+1}$ consider the return-time block for F_N'' containing it, and the return-time blocks just preceding and following this one. Be sure that $\mu(F_{N+1})$ is so small that the set of all points in these blocks has measure less than 2^{-N-11}. Modify each ϕ_i' to ϕ_i'', $i \leq N$ by making it the identity on these blocks. Now modify the sets F_i'', $i \leq N$ on these blocks so that all the new return-time blocks are still long enough for the collapsings on them to still be $m, 2^{-i} - 2^{-j}$-B&B, so that they are nested and all contain the new point in F_{N+1}. □

The following result now follows rather directly from this corollary. It is complex to state, but the basic picture is easy to understand.

Theorem A.1.13. *For each N and finite list of bounded rearrangements* $(\alpha, \phi_1), (\alpha, \phi_2), \ldots, (\alpha, \phi_N)$ *with* $J(\alpha, \phi_i) = 0$ *there is a value*

$$\delta_N((\alpha, \phi_1), \ldots, (\alpha, \phi_N))$$

so that if (α, ϕ_i) *is an infinite sequence of bounded rearrangements all with* J=0 *satisfying*

$$m(\alpha\phi_i, \alpha\phi_{i+1}) < \delta_i((\alpha, \phi_1), \ldots, (\alpha, \phi_i)),$$

then there is an m-nicely-blocked and bounded sequence of rearrangements (α, ϕ_i') *with*

$$\mu(\{x : \phi_i'(x) \neq \phi_i(x)\}) \leq 2^{-7}$$

for all i.

This will mean that there is a β *with* $m(\alpha\phi_i', \beta) \underset{i}{\to} 0$ *and hence that* $\alpha \overset{m}{\sim} \beta$. *Furthermore we will see that there is a* ψ *in the full-group with*

$$\|\alpha\phi_i, \beta\psi\|_1 \underset{i}{\to} 0$$

which is to say the $\alpha\phi_i$ are converging to an arrangement $\beta\psi$ that is m-equivalent to α.

Proof We will construct inductively using Lemma A.1.12. At stage N of the induction we will have constructed a sequence of rearrangements (α, ϕ_i^N), $i = 1, \ldots, N$. These will satisfy the hypotheses of Lemma A.1.12. The value δ_N will be the value δ_N of this Lemma. Supposing

$$m(\alpha\phi_N, \alpha\phi_{N+1}) < \delta_N,$$

let $\phi_{N+1}\phi_N^{-1}\phi_N$ be the "ϕ_N" of the lemma, as $(\alpha\phi_N, \phi_N^{-1}\phi_{N+1})$ and $(\alpha\phi'_N, \phi_N'^{-1}\phi_{N+1}\phi_N'^{-1}\phi_N)$ are identical in distribution. Lemma A.1.12 now tells us we can modify each of the ϕ_i^N to the new ϕ_i^{N+1} and add on the new term ϕ_{N+1}^{N+1}.

From Lemma A.1.12 we know that for each i,

$$\mu(\{x : \phi_i^N(x) \neq \phi_i^{N+1}(x)\}) \leq 2^{-N-9}$$

implying the full-group elements ϕ_i^N converge in N to the full-group element we call ϕ'_i with

$$\mu(\{x : \phi_i(x) \neq \phi'_i(x)\}) \leq \mu(\{x : \phi_i^i(x) \neq \phi_i(x)\}) + 2^{-i-8}.$$

Our inductive construction makes

$$\mu(\{x : \phi_i^i(x)\phi_i^{-1}(x) \neq \phi_{i-1}^{i-1}(x)\phi_{i-1}^{-1}(x)\}) \leq 2^{-i-9}$$

and so summing backwards to $i = 1$ we see that for all i

$$\mu(\{x : \phi'_i(x) \neq \phi_i(x)\}) \leq 2^{-7}.$$

(We will refine this estimate later.)

That the limit elements ϕ'_i form an m-nicely-B&B sequence of rearrangements is clear from the inductive construction.

Notice that for an m-nicely-blocked sequence of rearrangements (α, ϕ_i) the $\alpha\phi_i$ converge in m to some β.

We saw above that

$$\mu(\{x : \phi_i^i(x)\phi_i^{-1}(x) \neq \phi_{i-1}^{i-1}(x)\phi_{i-1}^{-1}(x)\}) \leq 2^{-i-9}$$

and so

$$\mu(\{\{x : \phi'_i(x)\phi_i^{-1}(x) \neq \phi'_{i-1}(x)\phi_{i-1}^{-1}(x)\}) \leq 2^{-i-6}$$

which is summable. This tells us that the $\phi_i'^{-1}\phi_i$ are L^1-Cauchy in the full-group, hence converging to some ψ. Thus $\alpha\phi_i = \alpha\phi'_i\phi_i'^{-1}\phi_i$ converges in $\mathscr{A}$ to the arrangement $\beta\psi$ finishing the result. □

The following corollary now indicates the link between 1^+-sizes and m-nicely-blocked sequences of rearrangements.

Corollary A.1.14. *Suppose m is a 1^+-size and $\alpha \overset{m}{\sim} \beta$. There is then an arrangement γ and full-group elements $\hat{\psi}_i$ and $\hat{\psi}'_i$, $i = 0, 1, \ldots$, so that the rearrangements*

$$(\alpha\hat{\psi}_0, \hat{\psi}_0^{-1}\hat{\psi}_i)$$

and

$$(\beta\hat{\psi}'_0, (\hat{\psi}'_0)^{-1}\hat{\psi}'_i)$$

are both m-nicely-blocked and bounded and both $m(\alpha\hat{\psi}_i, \gamma)$ and $m(\beta\hat{\psi}'_i, \gamma)$ tend to zero in i.

Proof We will begin by constructing sequences ψ_i and ψ'_i with

$$m(\alpha\psi_i, \psi_i^{-1}\psi_{i+1}) < \delta_i((\alpha\psi_0, \psi_0^{-1}\psi_1), \ldots, (\alpha\psi_0, \psi_0^{-1}\psi_i))$$

and

$$m(\beta\psi'_i, \psi_i'^{-1}\psi'_{i+1}) < \delta_i((\beta\psi'_0, \psi_0'^{-1}\psi'_1), \ldots, (\beta\psi'_0, \psi_0{}^{-1}\psi'_i))$$

and

$$m(\alpha\psi_i, \beta\psi'_i) \underset{i}{\rightarrow} 0.$$

Theorem A.1.13 now completes the result by giving $\hat{\psi}_i$ and $\bar{\psi}_i$ that are m-nicely B&B with both

$$\alpha\hat{\psi}_i \rightarrow \gamma \quad \text{and} \quad \beta\bar{\psi}_i \rightarrow \gamma'$$

and as $m(\alpha\psi_i, \beta\psi'_i) \rightarrow 0$, $\gamma = \gamma\psi$ for some ψ. Setting $\hat{\psi}'_i = \bar{\psi}_i\psi$ is all we need do.

To construct ψ_i and ψ'_i we will alternately add terms to each of the sequences ψ_i and ψ'_i. Moreover, at each stage all the full-group elements in both lists that we have already constructed will be slightly perturbed by right multiplication by an L^1-small full-group element. As we do this to all terms, we do not change the relations among them, and as these perturbations will converge in the full-group, the two sequences will converge to sequences in the full-group. The two full-group elements ψ_0 and ψ'_0 are in fact the total perturbations. In this process we will not index over this sequence of perturbations but will always just indicate ψ_i or ψ'_i, noting that they have been perturbed but have kept their names. To begin, let both ψ_0 and ψ'_0 be the identity (they will, as indicated above, change). The choice of ψ_1 is arbitrary, but choose it so that $(\alpha\psi_0, \psi_0^{-1}\psi_1)$ is bounded. Now using Theorem A.1.13 we get a value $\delta_1 = \delta_1(\alpha\psi_0, \psi_0^{-1}\psi_1)$. Choose ψ_1' so that both

(1) $m(\beta\psi_1', \alpha\psi_1) < \delta_1$ and

(2) $(\beta\psi_0', \psi_0'^{-1}\psi_1')$ is bounded.

Do this by first obtaining (1), then perturbing ψ_0' by at most ε_1 in L^1 to obtain (2) as well.

To construct ψ_2, notice we now can apply Theorem A.1.13 to the β side to get

$$\delta_1' = \delta_1'(\beta\psi_0', \psi_0'^{-1}\psi_1').$$

Choose ψ_2 so that both

(1) $m(\alpha\psi_2, \beta\psi_1') < \delta_1'$ and

(2) $(\alpha\psi_0, \psi_0^{-1}\psi_2)$ is bounded.

Do this by first obtaining (1), then perturbing both ψ_0 and ψ_1 by right multiplicating by a full-group element that is within some ε_2 in L^1 of the identity to obtain (2).

To construct ψ_2', apply Theorem A.1.13 to the α side to get

$$\delta_2 = \delta_2((\alpha\phi_0, \phi_0^{-1}\phi_1), (\alpha\phi_0, \phi_0^{-1}\phi_2)).$$

Choose ψ_2' so that both

(1) $m(\beta\psi_2', \alpha\psi_2) < \delta_2$ and

(2) $(\beta\psi_0', \psi_0'^{-1}\psi_2')$ is bounded.

Do this by first obtaining (1), then perturbing the list ψ_0', ψ_1' by right multiplying by a full-group element that is within some ε_2 in L^1 of the identity to obtain (2).

To finish we just continue this induction alternating sides and using successively the bounds δ_N of Theorem A.1.13. Making the ε_i summable will force convergence of the perturbations, and hence existence of the ψ_i and ψ_i' as described. □

This rather long effort has now established the role of nicely blocked sequences of rearrangements. The Fundamental Lemma (Lemma 4.8, page 100) of [43] actually describes an inductive step of precisely the same form as this result. The sequences of arrangements ($\alpha_0, \alpha_1, \ldots, \alpha_n$ and $\alpha_0', \alpha_1', \ldots, \alpha_m'$) constructed there are not directly assumed to be m-nicely-B&B but the mass of structure in the definition of an (n, m)-approximation says we could in fact collapse all the terms and get a sequence that is.

This observation is not going to matter much to us as boundedness is all that will really matter in our discussion of m-f.d. actions. We point this out here just to indicate the parallelism.

We proceed to construct an r-size associated with every 1-size which, for 1^+-sizes will have precisely the same equivalence classes, and universally the same finitely determined classes. Both these results will be based on the fact that for a sequence of rearrangements $\alpha\phi_i \underset{i}{\to} \beta$ and $\beta\phi_i^{-1} \underset{i}{\to} \alpha$ in L^1 which are bounded in both directions, convergence for the 1-size and the corresponding r-size will be equivalent.

Let m be a 1-size. Our first simple task is to bound m. For a permutation π, define

$$m^1(\phi) = \min(m(\pi), 1).$$

Lemma A.1.15. *If m is a 1-size, then m^1 as defined above is as well. Moreover, this new 1-size has the same equivalence classes as m.*

Proof It is quite simple to check the six conditions of [43]. Notice that using m^1 the requirement that $m(\alpha_2, \alpha_3) < 1$ in axiom (vi) can be completed omitted, i.e. one has:

(iv) given any $\varepsilon > 0$ there is a $\delta > 0$ so that if $m^1(\alpha_1, \alpha_2) < \delta$ then for any α_3,

$$m^1(\alpha_1, \alpha_3) \leq m^1(\alpha_2, \alpha_3) + \varepsilon.$$

To see that the equivalence classes are unchanged, just remember that for $\alpha_1 \overset{m}{\sim} \alpha_2$ means there are full-group elements ϕ_i with

$$m(\alpha_1\phi_i, \alpha_2) \underset{i}{\to} 0.$$

We observe that $m^1(\alpha, \beta) = \min(m(\alpha, \beta), 1)$. Thus $m(\alpha_1\phi_i, \alpha_2) \underset{i}{\to} 0$ is equivalent to $m^1(\alpha_1\phi_i, \alpha_2) \underset{i}{\to} 0$. Thus the two equivalence relations agree. □

Our next step is to replace m^1 with a metric. To do this we must leave behind permutations. We use a standard trick for creating a metric.

For any arrangement α_1 define a function:

$$F_{\alpha_1} : \mathscr{A} \to [0, 1] \quad \text{by} \quad F_{\alpha_1}(\alpha_2) = m^1(\alpha_1, \alpha_2).$$

Set

$$m^2(\alpha_1, \alpha_2) = \|F_{\alpha_1}, F_{\alpha_2}\|_\infty.$$

It is easy to see that m^2 is a metric. The only non-obvious point is that the map $\alpha_1 \to F_{\alpha_1}$ is 1-1. To see this just note that the only zero of this function is at α_1.

Lemma A.1.16. *Given any $\varepsilon > 0$ there is a $\delta > 0$ so that:*

(*a*) *If $m^1(\alpha_1, \alpha_2) < \delta$ then $m^2(\alpha_1, \alpha_2) < \varepsilon$ and*
(*b*) *if $m^2(\alpha_1, \alpha_2) < \delta$ then $m^1(\alpha_1, \alpha_2) < \varepsilon$.*

Proof Statement (b) is trivial as

$$m^2(\alpha_1, \alpha_2) = \sup_{\alpha_3}(|m^1(\alpha_1, \alpha_3) - m^1(\alpha_2, \alpha_3)|) \geq m^1(\alpha_1, \alpha_2).$$

For (a), using (iv) stated above for m^1, if $m^1(\alpha_1, \alpha_2) < \delta$ then for all α_3,

$$|m(\alpha_1, \alpha_3) - m(\alpha_2, \alpha_3)| < \varepsilon$$

which is what we want. □

Lemma A.1.17. *For m a 1-size, $\alpha_1 \overset{m}{\sim} \alpha_2$ if and only if α_2 is in the m^2-closure of the full-group orbit of α_1.*

Proof To say $m(\alpha_1\phi_i, \alpha_2) \underset{i}{\to} 0$, as we just saw in Lemma A.1.16 is equivalent to $m^2(\alpha_1\phi_i, \alpha_2) \underset{i}{\to} 0$. □

We need to take two more steps in modifying m to become an r-size. At present m^2 satisfies Axioms 1 and 2 of an r-size (Axiom 2, that $(\Gamma, m^2_\alpha) \overset{\text{id}}{\to} (\Gamma, \|\cdot, \cdot\|^\alpha_w)$ is uniformly continuous follows from Axiom (iv) of a 1-size, Lemma A.1.16 and a bit of thought). As we indicated earlier Axiom 3 is not clearly associated with the conditions of a 1-size and acquiring it will take a little two-step.

Definition A.1.18. *For m a 1-size let*

$$m^3(\alpha_1, \alpha_2) = \inf(m^2(\alpha_1, \alpha_2\phi) + \mu(\{x : \phi(x) \neq x\}) : \phi \in \text{FG}(\mathcal{O})).$$

Note: we use $\mu(\{x : \phi(x) \neq x\})$ rather than $\|\phi, \text{id}\|^\alpha_s$ as it has better equivariance properties. In particular relative to the metric $\mu(\{x : \phi_1(x) \neq \phi_2(x)\})$ the full-group is a Polish metric space and this metric is invariant under both right and left composition.

Lemma A.1.19. *For m a 1-size, m^3 is a metric and for any arrangement α, the m-orbit closure of $\{\alpha\phi\}$ is the same as the m^3-orbit closure. Moreover m^3 satisfies Axiom 2 of an r-size.*

Proof To check the triangle inequality, suppose $m^3(\alpha_1, \alpha_2) = a$ and $m^3(\alpha_2, \alpha_3) = b$. For any $\varepsilon > 0$ there exist ϕ_1 and ϕ_2 with

$$\begin{aligned} a \leq& m^2(\alpha_1, \alpha_2\phi_1) + \mu(\{x : \phi_1(x) \neq x\}) + \varepsilon, \\ b \leq& m^2(\alpha_2, \alpha_2\phi_2) + \mu(\{x : \phi_2(x) \neq x\}) + \varepsilon \\ =& m^2(\alpha_2\phi_1, \alpha_3\phi_2\phi_1) + \mu(\{x : \phi_2\phi_1(x) \neq \phi_1(x)\}) + \varepsilon. \end{aligned}$$

Thus

$$\begin{aligned} m^3(\alpha_1, \alpha_3) \leq& m^2(\alpha_1, \alpha_3\phi_2\phi_1) + \mu(\{x : \phi_2\phi_1(x) \neq x\}) \\ \leq& a + b + 2\varepsilon. \end{aligned}$$

To see that the orbit closures agree, notice that if $m^3(\alpha_1, \alpha_2\phi) < \varepsilon$ then there is a ϕ' with $m^2(\alpha_1, \alpha_2\phi\phi') + \mu(\{x : \phi'(x) \neq x\}) < \varepsilon$ and hence $m^2(\alpha_1, \alpha_2\phi\phi') < \varepsilon$ and the m^2 orbit closure contains the m^3. The other direction is trivial.

As m^2 satisfies Axiom 2 of an r-size, and $\mu(\{x : \phi(x) \neq x\})$ is an r-size, m^3 satisfies Axiom 2 as well. □

Definition A.1.20. *Let*

$$m^4(\alpha, \phi) = \limsup_{\|(\alpha,\phi),(\alpha',\phi')\|_* \to 0} m^3(\alpha', \alpha'\phi').$$

Our next task is to show that m^4 is an r-size.

Lemma A.1.21. *For m a 1-size and α an arrangement $m^4_\alpha(\phi_1, \phi_2) = m^4(\alpha\phi_1, \phi_1^{-1}\phi_2)$ is a pseudometric on the full-group.*

Proof This amounts to verifying the triangle inequality, i.e. given α, ϕ_1 and ϕ_2 that

$$m^4(\alpha, \phi_2) \leq m^4(\alpha, \phi_1) + m^4(\alpha\phi_1, \phi_1^{-1}\phi_2).$$

This is a job for a copying lemma.

By Corollary 4.0.9, for any $\delta > 0$ there is a $\delta_1 > 0$ so that if

$$\|(\alpha, \phi_2), (\alpha', \phi_1')\|_* < \delta_1,$$

then there is a ϕ_1' in the full-group of $T^{\alpha'}$ with both

$$\|(\alpha, \phi_1), (\alpha', \phi_1')\|_* < \delta \quad \text{and} \quad \|(\alpha\phi_1, \phi_1^{-1}\phi_2), (\alpha', \alpha'\phi_1'^{-1}\phi_2')\|_* < \delta.$$

Note: This is stretching Corollary 4.0.9 a bit. To see that this stretch is acceptable notice that as

$$\|(T^{\alpha}, g_{(\alpha,\phi_1)} \vee g_{(\alpha,\phi_2)}), (T^{\alpha'}, g_{(\alpha',\phi_1')} \vee g_{(\alpha',\phi_2')})\|_* \to 0$$

we will have both

$$\|(\alpha, \phi_1), (\alpha', \phi_1')\|_* \to 0 \qquad \text{and}$$
$$\|(\alpha\phi_1, \phi_1^{-1}\phi_2), (\alpha', \alpha'\phi_1'^{-1}\phi_2')\|_* \to 0.$$

Thus use finite partitions that approximate the two possibly infinite partitions $g_{(\alpha,\phi_2)}$ and $g_{(\alpha',\phi_2')}$ as Q and Q_1 in Corollary 4.0.9.

Continuing, for any $\varepsilon > 0$ we can choose δ and δ_1 small enough that

$$m^4(\alpha, \phi_2) \le m^3(\alpha', \alpha'\phi_2') + \varepsilon,$$
$$m^3(\alpha', \alpha'\phi_1') \le m^4(\alpha, \phi_1) + \varepsilon, \text{ and}$$
$$m^3(\alpha'\phi_1', \phi_1'^{-1}\phi_2') \le m^4(\alpha\phi_1, \phi_1^{-1}\phi_2) + \varepsilon.$$

We conclude

$$m^4(\alpha, \phi_2) \le m^4(\alpha, \phi_1) + m^4(\alpha\phi_1, \phi_1^{-1}\phi_2) + 3\varepsilon$$

to complete the result. □

Theorem A.1.22. *For m a 1-size, m^4 is an r-size.*

Proof Lemma A.1.21 says we only need to verify Axioms 2 and 3. Lemma A.1.19 tells us m^3 satisfies Axiom 2 and as

$$m^4(\alpha, \phi) \ge m^3(\alpha, \phi)$$

we conclude m^4 does as well. The definition of m^4 is such as to make Axiom 3 direct. □

We will complete our work now by showing that for any 1-size, if $\alpha_1 \overset{m}{\sim} \alpha_2$ by an m-nicely-blocked sequence of rearrangements then $\alpha_1 \overset{m^4}{\sim} \alpha_2$.

Suppose (α, ϕ) is B&B over a set F and hence r_F is bounded. For each $x \in F$ let π_x^F be the permutation of $(0, 1, \ldots, r_F(x) - 1)$ with

$$\phi(T_j^{\alpha}(x)) = T^{\alpha}_{\phi_x^F(j)}(x),\ 0 \le j < r_F(x).$$

This is a finite list of permutations. Add on to this all identity permutations of finite blocks $(0, \ldots, N-1)$ and call the resulting collection of permutations $\Pi(\phi, F)$. This is an infinite set of permutations only because we included all the identities.

Lemma A.1.23. *Suppose (α, ϕ) is blocked and bounded over a set F, and for some N, $r_F(x) \geq N$ for all $x \in F$. Then for any $\varepsilon > 0$ there is a δ so that for any (α', ϕ') with*

$$\|(\alpha, \phi), (\alpha', \phi')\|_* < \delta$$

there is a ϕ'' with $\mu(\{x : \phi'(x') \neq \phi''(x')\}) < \varepsilon$ and (α', ϕ'') is blocked and bounded over some set F' with

1. *$\Pi(\phi'', F') \subseteq \Pi(\phi, F)$ and*
2. *$r_{F'}(x') \geq N$ for all $x' \in F'$.*

Proof As r_F is bounded let $r_F \leq B$ and choose $4M = [10B/\varepsilon]+1$. Choose δ so small that knowing

$$\|(\alpha, \phi), (\alpha', \phi')\|_* < \delta$$

implies that for all but $\varepsilon/10$ in measure of the $x' \in X'$ there is an $x \in X$ with

$$f^{\alpha,\phi}(T_j^{\alpha}(x)) = f^{\alpha',\phi'}(T_j^{\alpha'}(x')), \quad 0 \leq j < M.$$

Call a choice for this point $x(x')$. We can and do choose the point $x(x')$ to depend only on the list of values $f^{\alpha',\phi'}(x')$, $0 \leq j < M$. Hence there are only finitely many points of the form $x(x')$. Let Q be the finite partition of X' into sets according to the value $f^{\alpha',\phi'}(x)$ if it is $\leq B$ in absolute value, and the complimentary set. Construct a Rokhlin tower for $T^{\alpha'}$ with base C and height M, covering all but $\varepsilon/10$ of X' with

$$C \perp \bigvee_{j=0}^{M-1} T_{-j}^{\alpha'}(Q).$$

Let $C_0 \subseteq C$ consist of those x' for which there is **no** value $x(x')$. Hence $\mu(C_0) \leq \frac{\varepsilon}{10}\mu(C)$.

Partition the remainder of C according to $\vee_{j=0}^{M-1} T_{-j}^{\alpha'}(Q)$ which is the same as to say, according to the value $x(x')$. Call these sets $C_1, C_2, \ldots, C_k$. Let x_i be the value $x(C_i)$, $i \geq 1$.

For each x_i let

$$0 \leq j_1^i < j_2^i < \cdots < j_{L(i)}^i < M$$

be those indices with

$$T_{j_t^i}^{\alpha}(x_i) \in F.$$

Any two successive values $j_t^i < j_{t+1}^i$ are at least N and at most B apart.

To start a definition of F', put

$$\bigcup_{i=1}^{K} \bigcup_{t=2}^{l(i)-1} T^{\alpha'}_{j^i_t}(C_i)$$

in F'.

To start a definition of ϕ'', for each $x' \in C_i$ and each block of points

$$(T^{\alpha'}_{j^i_t}(x'), \ldots, T_{j^i_{t+1}-1}(x')), \quad 2 \leq t < L(i) - 1$$

just set $\phi'' = \phi$. Thus, as this is a block between two occurrences of F' and the functions $f^{\alpha',\phi''}(x')$ and $f^{\alpha,\phi}(x_i)$ agree on $(0, M-1)$, we have

$$\pi^{\phi''}_{T^{\alpha'}_{j^i_t}(x')} = \pi^{\phi}_{T^{\alpha}_{j^i_t(x_i)}}.$$

For all other points x', those in the tower over C_0 or outside of the tower, or below level j^i_2 or above level $j^i_{L(i)-1} - 1$ in the tower over C_i, set $\phi''(x') = x'$. Notice this is a full-group element. Moreover, notice that the set on which we just defined $\phi'' = \text{id}$ always occurs in an orbit in blocks at least $2N$-long. In each such block then we can select a list of points to add to F' so that $r_{F'}$ is bounded, but is always $\geq N$. On each such block, the permutation induced by ϕ'' will just be the identity. Both conditions (1) and (2) are now evident. □

Corollary A.1.24. *Suppose m is a 1-size. Given any $\varepsilon > 0$ there is a $\delta > 0$ so that if (α, ϕ) is bounded, $J(\alpha, \phi) = 0$ and $m(\alpha, \alpha\phi) < \delta$. Then $m^4(\alpha, \phi) < \varepsilon$.*

Proof Choose ε_1 so that if $m(\alpha, \phi) < \varepsilon_1$ then $m^3(\alpha, \phi) < \varepsilon$. Choose δ so that if $m(\alpha, \alpha\phi) < \delta$ and bounded with $J(\alpha, \phi) = 0$ then for all δ_1 and K there are m, ε_1-nicely blocked ε, δ_1, K-collapsings $(\alpha\phi_1)$ of $\alpha\phi$ onto α with $\mu(\{x : \phi_1(x) \neq \phi(x)\}) < \delta_1$. Arguing analogously to Lemma A.1.23 for any (α', ϕ') close enough in distribution to (α, ϕ) we will be able to copy the collapsing $\alpha\phi_1$ into (α', ϕ') as an $\varepsilon_1, \delta_1, K$-collapsing with $\mu_1(\{x' : \phi'_1(x') \neq \phi'(x')\}) < 2\delta_1$. In particular the rearrangement (α', ϕ'_1) will be m, ε_1-nicely-blocked and hence $m(\alpha', \phi'_1) < \varepsilon_1$ and $m^3(\alpha', \phi') \leq \varepsilon$. But of course this says $m^4(\alpha, \phi) < \varepsilon$. □

Theorem A.1.25. *Suppose m is a 1-size. For any α_1 and α_2 for which there are ϕ_i, $m(\alpha_1\phi_i, \alpha_2) \underset{i}{\rightarrow} 0$ and all (α_1, ϕ_i) and (α_2, ϕ_i^{-1}) are bounded we will conclude $\alpha_1 \overset{m^4}{\sim} \alpha_1$. In particular, if m is a 1^+-size then whenever $\alpha_1 \overset{m}{\sim} \alpha_2$, we also have $\alpha_1 \overset{m^4}{\sim} \alpha_2$ and any T which is m-f.d. is m^4-f.d.*

Proof Corollary A.1.24 tells us that both (α_1, ϕ_i) and (α_2, ϕ_i^{-1}) must be m^4-Cauchy, completing the result. To finish the other observations, notice that any nicely-blocked sequence of rearrangements is bounded in both the forward and backward directions. Corollary A.1.14 tells us that we can interpose a γ between α_1 and α_2 that will be m^4-equivalent to both of them. That the notion of m-f.d. implies m^4-f.d. is simply that the joining demanded in the definition of m-f.d. in [43] has to be a bounded rearrangment and hence m^4-small. □

We now want to see that if $\alpha_1 \overset{m^4}{\sim} \alpha_2$ then $\alpha_1 \overset{m}{\sim} \alpha_2$. This may seem trivial now as we know that any m^4-Cauchy sequences of rearrangements will be m^3-Cauchy. The problem is that we do not know that just because (α_1, ϕ_i) is m-Cauchy, and $\alpha_1\phi_i \to \alpha_2$ in L^1 that then $m(\alpha_1\phi_i, \alpha_2) \to 0$. The way we handle this issue is much like what we already did for the other direction; we introduce convergence criterion.

Definition A.1.26. *Suppose (X, d) is a metric space and $m : X \times X \to \mathbb{R}^+$. We say a sequence of functions*

$$\delta_n : X^n \to \mathbb{R}^+$$

is a convergence criterion for the function m if for any sequence of values $\{x_n\}$ in X with

$$m(x_i, x_{i+1}) \leq \delta_i(x_1, \ldots, x_i)$$

there is a (necessarily unique) point $x \in X$ with

$$d(x_i, x) \underset{i}{\to} 0.$$

Lemma A.1.27. *For m a 1-size, and $\langle\alpha\rangle_m$ the m-equivalence class of an arrangement α, we know m^2 is a metric on $\langle\alpha\rangle_m$. Relative to the metric space $(\langle\alpha\rangle_m, m^2)$, there are convergence criteria for both m and m^2.*

Proof This construction just abstracts what is done in both Theorem 3.1 and Lemma 4.8 of [43]. To construct the convergence criterion for m, select sets $A_1, A_2, \ldots$ and bounds $K_1 < K_2 < \cdots$ inductively with

(1) $\mu(A_N) > 1 - 2^{-(N+2)}$ for all N, and for each $x \in A_N$ there is a list of blocks, $(n_{i,N}, m_{i,N})$ with

$$n_{i,N}(x) \leq -N < 0 < N \leq m_{i,N}(x)$$

so that

(2) $[-N,N] \subseteq f_x^{\alpha_1,\alpha_N}([-n_{i,M}(x), m_{i,n}(x)]) \subseteq [-K_N, K_N]$ and

(3) $m(\pi_{f_x^{\alpha_i,\alpha_N},(n_{i,N}(x),m_{i,N}(x))}) < m(\alpha_i,\alpha_N) + 2^{-(N+1)}$.

That $m(\alpha_1,\alpha_N) = m(f_x^{\alpha_1,\alpha_N})$ μ-a.s. and that $f_x^{\alpha_1,\alpha_N}$ is a bijection tell us we can select such bounds K_N and sets A_N.

Choose $\delta_N(\alpha_1,\ldots,\alpha_N)$ so small that if

$$m(\alpha_N, \alpha_N + 1) < \delta_N(\alpha_1,\ldots,\alpha_N)$$

then for all $i \leq N$,

(a) $m(\alpha_i,\alpha_{N+1}) < m(\alpha_i,\alpha_N) + 2^{-(N+1)}$ and

(b) $\mu(\{x : f_x^{\alpha_N,\alpha_{N+1}}|_{[-K_N,K_N]} = \mathrm{id}\}) > 1 - 2^{-(N+2)}$.

That we can do this is a consequence of Lemma 2.6 concerning m_0 and Axiom (v) both of [43], which requires that all 1-sizes essentially dominate m_0.

Suppose $\alpha_1, \alpha_2, \ldots$ satisfy

$$m(\alpha_{i+1},\alpha_i) < \delta_i(\alpha_1,\ldots,\alpha_i).$$

Let $K_1, K_2, \ldots$ and $A_1, A_2, \ldots$ be the associated sets and bounds. Let

$$\hat{A}_j = \{x : x \in A_N, N \geq j \text{ and } f_x^{\alpha_N,\alpha_{N+1}}|_{[-K_N,K_N]} = \mathrm{id}, N \geq j\}.$$

Notice that $\mu(\hat{A}_j) \geq 1 - 2^{-j}$. For $x \in \hat{A}_j$, for each $N \geq j$ and $i < N$ we will have values

$$n_{i,N}(x) < -N < 0 < N < m_{i,m}(x),$$

$$[-N,N] \subseteq f_x^{\alpha_i,\alpha_N}([n_{i,N}(x), m_{i,N}(x)]) \subseteq [-K_N, K_N]$$

and further, for all $k \geq N$,

$$f_x^{\alpha_k,\alpha_{k+1}}|_{[-K_k,K_k]} = \mathrm{id}.$$

As the K_k's increase, by composing we see

$$f_x^{\alpha_N,\alpha_{k+1}}|_{[-K_N,K_N]} = \mathrm{id}$$

and so for all $j \geq N$,

$$f_x^{\alpha_i,\alpha_j}|_{[n_{i,N}(x),m_{i,N}(x)]} = f_x^{\alpha_i,\alpha_N}|_{[n_{i,N}(x),m_{i,N}(x)]}.$$

As $[-N,N] \subseteq [n_{i,N}(x), m_{i,N}(x)]$ we conclude that there is a 1-1 function $f_{x,i}$ with

$$f_x^{\alpha_i,\alpha_j} \underset{j}{\rightarrow} f_{x,i}.$$

As $[-N, N] \subseteq f_x^{\alpha_i,\alpha_N}([n_{i,N}(x), m_{i,N}(x)])$ we conclude that $f_{x,i}$ is a bijection. Moreover it is a calculation to see that

$$f_x^{\alpha_i,\alpha_j} f_{x,j} = f_{x,i}.$$

Hence defining an arrangement

$$\beta(x, T_j^{\alpha_1}(x)) = f_{x,1}(j),$$

we have

$$\beta(x, T_j^{\alpha_i}(x)) = f_{x,i}(j).$$

It remains only to show that $m(\alpha_i, \beta) \to 0$ which is equivalent to $m^1(\alpha_i, \beta) \to 0$.

To see this just notice that for $x \in \hat{A}_j$, $n \geq j$ and $i < N$

$$m(\phi_{f_x^{\alpha_i,\alpha_N},(n_{i,N}(x),m_{i,N}(x))}) = m(\phi_{f_x^{\alpha_i,\beta},(n_{i,N}(x),m_{i,N}(x))})$$
$$< m(\alpha_i, \alpha_N) + 2^{-(n+1)} < 2^{-i}.$$

Letting $N \to \infty$ we conclude that $m(f_x^{\alpha_i,\beta}) \leq 2^{-i}$ and we are done. □

As $m^1(\alpha_1, \alpha_2) = m(\alpha_1, \alpha_2)$ once $m(\alpha_1, \alpha_2) \leq 1$ a convergence criterion for m will be one for m^1 as well, and as m^1 and m^2 are uniformly related, there is a convergence criteria for m^2.

Lemma A.1.28. *For m a 1-size and $\langle \alpha \rangle_m$ the m-equivalence class of α, relative to the metric space $(\langle \alpha \rangle_\alpha, m^3)$ there is a convergence criterion for m^3.*

Proof To construct a convergence criterion for m^3 select inductively, for a list $\alpha_1, \ldots, \alpha_N$, full-group elements $\phi_1, \ldots, \phi_N$ so that

$$m^3(\alpha_i, \alpha_{i+1}) \leq 2(m^2(\alpha_i\phi_i, \alpha_{i+1}\phi_{i+1})) + \mu(\{x : \phi_i(x) \neq \phi_2(x)\}).$$

Let δ_N be the convergence criterion for m^2 and now define

$$\delta'_N(\alpha_1, \alpha_2, \ldots, \alpha_N) = \min(\delta_N(\alpha_1\phi_1, \alpha_2\phi_2, \ldots, \alpha_N\phi_N), 2^{-N}).$$

Suppose now that α_i satisfies

$$m^3(\alpha_i, \alpha_{i+1}) < \delta'_N(\alpha_1, \ldots, \alpha_N).$$

Then both

$$m^2(\alpha_i\phi_i, \alpha_{i+1}\phi_{i+1}) < \delta_N(\alpha_1\phi_1, \ldots, \alpha_i\phi_i) \quad \text{and}$$

$$\mu(\{x : \phi_i(x) \neq \phi_{i+1}(x)\}) < 2^{-i}.$$

We conclude that there must be a $\beta \in \langle\alpha\rangle_m$ with

$$m^2(\alpha_i\phi_i, \beta) \underset{i}{\rightarrow} 0$$

and a ϕ with

$$\mu(\{x : \phi_i(x) \neq \phi(x)\}) \underset{i}{\rightarrow} 0.$$

Hence $m^3(\alpha_i\phi, \beta) \underset{i}{\rightarrow} 0$ which is the same as $m^3(\alpha_i, \beta\phi) \underset{i}{\rightarrow} 0$. □

Theorem A.1.29. *For m a 1-size, if $\alpha \overset{m^4}{\sim} \beta$ then $\alpha \overset{m}{\sim} \beta$.*

Proof The "trick" here is exactly the same as that in Corollary A.1.14. We give a brief sketch. Restrict the discussion to the metric space $(\langle\alpha\rangle_{m^4}, m^4)$. Remember m^4 is the "geodesic" distance, the infimum over sequences of full-group elements leading from one to the other in both directions.

What we construct now are two sequences of full-group elements ψ_i and ψ_i' so that (α, ψ_i) and (β, ψ_i') both satisfy the convergence criterion for m^3 and

$$\|\alpha\psi_i, \beta\psi_i'\|_1 \underset{i}{\rightarrow} 0.$$

This of course completes the result as both $\alpha\psi_i$ and $\beta\psi_i'$ will be converging to a common γ to which they are both m^3- and hence m-equivalent.

The construction follows the same lines as that in Corollary A.1.14 by successively pulling in from one side and then the other by full-group elements close enough in $m^4 \leq m^3$ to be sure to obtain the next term of the convergence criterion on the other side at the next step. □

A.2 p-sizes

In a preliminary version of our work on sizes for discrete amenable group actions we put forward an axiomatization that lies between what we have actually set out here and the axiomatization we used in [25]. What we will do in this section is to state that axiomatization, show that a "size" in those terms gives rise to a size in satisfying our axioms here having precisely the same equivalence classes. We then will present the axiomatization used in [25] and show that a size in those terms (which we call a p-size) gives rise to a size in terms of our older axiomatization for discrete amenable groups.

Here is that preliminary axiomatization: a size m, as defined in a preliminary version of this work, is a function m from the space of G-rearrangements to $\mathbb{R}^+$ that satisfies the following five axioms:

Axiom B1. *Given $\varepsilon > 0$ there exists $\delta > 0$ such that if $\|\phi\|_{\text{p.w.}} < \delta$ then $m(\alpha, \phi) < \varepsilon$.*

Axiom B2. *Given $\varepsilon > 0$ there exists $\delta > 0$ such that if $m(\alpha, \phi) < \delta$ then $\|\alpha, \alpha\phi\|_{\text{p.w.}} < \varepsilon$.*

Axiom B3. *Given $\varepsilon > 0$ there exists $\delta > 0$ such that if $m(\alpha, \phi) < \delta$ then $m(\alpha\phi, \phi^{-1}) < \varepsilon$.*

Axiom B4. *Given $\varepsilon > 0$ there exists $\delta > 0$ such that if $m(\alpha\phi, \psi) < \delta$ then $m(\alpha, \phi\psi) < m(\alpha, \phi) + \varepsilon$.*

Axiom B5. *Given $\varepsilon > 0$ and rearrangement (α, ϕ), there exists $\delta = \delta(\varepsilon, \alpha, \phi) > 0$ such that if $\|(\alpha, \phi), (\beta, \psi)\|_* < \delta$ then $m(\beta, \psi) < m(\alpha, \phi) + \varepsilon$.*

Suppose m is a size function, in the sense that it satisfies the above five (old) axioms. We will construct a new size function m^2 which satisfies the current three axioms, giving the same full-group topology, and the same completion. That is to say, the m-Cauchy sequences in the full-group are exactly the m^2-Cauchy sequences in the full-group. The two steps we take are precisely analogous to those taking m^2 to m^4 in our work on 1-sizes. It is worth pointing out here that Axiom B1 and Theorem 4.0.2 tell us that we have the analogue of a 1^+-size in that we can always perturb elements of the full-group by a small amount in m to be bounded. Note that the size function m is a conjugacy invariant, meaning that

$$m(\alpha, \phi) = m(\alpha\psi, \psi^{-1}\phi\psi).$$

To each arrangement α and full-group element ϕ_0, associate a function

$$k^{\alpha}_{\phi_0} : \Gamma \to \mathbb{R}^+$$

by letting

$$k^{\alpha}_{\phi_0}(\phi) = m(\alpha\phi, \phi^{-1}\phi_0).$$

Axioms B1 and B3 imply that $k^{\alpha}_{\phi_0}$ is uniformly continuous with respect to the L^1 topology on the full-group.

For ϕ_1 and ϕ_2 in the full-group, define

$$m_\alpha(\phi_1, \phi_2) = m(\alpha\phi_1, \phi_1^{-1}\phi_2),$$

and define

$$m^1(\alpha, \phi) = \|k^\alpha_{\mathrm{id}}, k^\alpha_\phi\|_{\sup}$$

and so

$$m^1_\alpha(\phi_1, \phi_2) = \|k^\alpha_{\phi_1}, k^\alpha_{\phi_2}\|_{\sup}.$$

Clearly, $m^1_\alpha(\cdot, \cdot)$ is a pseudo-metric on the full-group. Axioms B1, B3 and B4 imply that m_α and m^1_α give rise to equivalent topologies on the full group. In particular, any sequence that is m_α-Cauchy is also m^1_α-Cauchy, and vice versa.

Axiom B2 tells us that the identity map on the full-group is uniformly continuous from the m^1 to the L^1-metric.

All that remains is to obtain Axiom 3 of an r-size, i.e. upper semi-continuity with respect to distribution. We do this in exactly the same way as for 1-sizes.

Define

$$m^2 : \{G\text{-rearrangements}\} \to \mathbb{R}^+$$

by letting

$$m^2(\alpha, \phi) = \limsup_{\|(\alpha,\phi),(\beta,\psi)\|_* \to 0} m^1(\beta, \psi).$$

Also define

$$m^2_\alpha(\phi_1, \phi_2) = m^2(\alpha\phi_1, \phi_1^{-1}\phi_2).$$

That m^2 is a pseudometric follows exactly as in Lemma A.1.21 relating m^3 to m^4 for 1-sizes, as the copying lemma (Corollary 4.0.9) applies in precisely the same way.

Lemma A.2.1. *m^2_α and m^1_α are equivalent metrics on the full-group. That is to say, given any $\varepsilon > 0$ there is a δ so that*

(1) *if $m^1_\alpha(\phi_1, \phi_2) < \delta$ then $m^2_\alpha(\phi_1, \phi_2) < \varepsilon$ and*

(2) *if $m^2_\alpha(\phi_1, \phi_2) < \delta$ then $m^1_\alpha(\phi_1, \phi_2) < \varepsilon$.*

Proof

To see this just note that one can rewrite Axiom B5 as:
Given any (α, ϕ),

$$\limsup_{\|(\alpha,\phi),(\beta,\psi)\|_* \to 0} m(\beta, \psi) \le m(\alpha, \phi).$$

As we know that m and m^1 satisfy (1) and (2) above we conclude the lemma. □

Notice here that we could have used less than Axiom B5. To conclude that m^2 was equivalent to m all we really needed is the following.

Axiom B5′. *Given any $\varepsilon > 0$ there exists δ so that for any (α, ϕ) with $m(\alpha, \phi) < \delta$ then*

$$\limsup_{\|(\alpha,\phi),(\beta,\psi)\|_{\bullet}\to 0} m(\beta, \psi) < \varepsilon.$$

In [25], we presented a framework for restricted orbit equivalence for actions of $\mathbb{Z}^d$. A size function was initially defined on permutations of boxes in $\mathbb{Z}^d$. Such a size was required to satisfy six different axioms. This function was then extended to be defined on rearrangements, (α, ϕ), in many ways analogous to what happened for a 1-size. We will avoid all the problems we had in dealing with 1-sizes here in that rather than "collapsing" by "pushing-together" (something that does not make much sense beyond $\mathbb{Z}^1$), we "fill in", analogous to what we did in the proof of Theorem 4.0.2.

Let Π_n be the set of all permutations of the box B_n (in $\mathbb{Z}^d$), and let $\Pi = \cup_n \Pi_n$. According to [25], a p-size (p signifying "permutation") is a function $m : \Pi \to [0, 1]$ that satisfies the following axioms.

Axiom C1. (*Continuity near the identity*)

(a) *$m(id) = 0$ for all identity permutations, and*
(b) *for all $\varepsilon > 0$ there is an N and $\delta > 0$ so that for $n \geq N, \pi \in \Pi_n$, and if*

$$\frac{|\{\vec{v} \in B_n; \pi(\vec{v}) \neq \vec{v}\}|}{|B_n|} < \delta,$$

then $m(\pi) < \varepsilon$.

Axiom C2. ("*Lumpiness*" *of small permutations*)

For all N_0 and $\varepsilon > 0$, there is an N_1 and $\delta > 0$ so that if $n \geq N_1$ and $\pi \in \Pi_n$ with $m(\pi) < \delta$, then

$$\frac{|\{\vec{v} \in B_n;\ \text{for some } \vec{v}_0 \in B_{N_0}, \pi(\vec{v} + \vec{v}_0) \neq \pi(\vec{v}) + \vec{v}_0\}|}{|B_n|} < \varepsilon.$$

To make $\pi(\vec{v} + \vec{v}_0)$ well defined we extend π outside B_n as the identity. In such contexts we will always make this choice.

Axioms C1 and C2 are partial converses to one another. Axiom C1 tells us that if a permutation moves very few points, then it must be small. Axiom C2 tells us that if a permutation is small it must move most points in large rigid lumps.

Axiom C3. (*Continuity of inverses near zero*)
For all $\varepsilon > 0$ *there is an* N *and* $\delta > 0$ *so that for* $n \geq N$ *and* $\pi \in \Pi_n$ *with* $m(\pi) < \delta$, *then* $m(\pi^{-1}) < \varepsilon$.

Axiom C4. (*Continuity of composition near zero*)
For all $\varepsilon > 0$ *there is an* N *and* δ *so that for* $n \geq N$ *and* $\pi_1, \pi_2 \in \Pi_n$, *if* $m(\pi_2) < \delta$, *then*

$$m(\pi_2\pi_1) < m(\pi_1) + \varepsilon.$$

Axioms C3 and C4 simply tell us the kind of uniform continuity we will want of the group operations within each Π_n. There are two other natural operations within Π and that is the concatenation of permutations on disjoint blocks, and the rearranging of blocks left invariant by a permutation. One of these builds a new permutation on a larger block from smaller ones, the other changes a given permutation to a new one. Our last two axioms concern continuity of such operations.

Axiom C5. (*Continuity of concatenation near zero*)
For all $\varepsilon > 0$ *there is an* N_0 *and* $\delta > 0$ *so that if* $n_1, n_2, \ldots, n_k > N_0$ *and if* $\vec{v}_1, \vec{v}_2, \ldots, \vec{v}_k \in \mathbb{Z}^d$ *are such that*:

(a) $A_i = B_{n_i} + \vec{v}_i$ *are disjoint and for some* n;

(b) $\cup_{i=1}^k A_i \subseteq B_n$ *and* $\dfrac{|\bigcup_{i=1}^k A_i|}{|B_n|} > 1 - \delta$:

(c) $\pi \in \Pi_n$ *is such that for all* i, $\pi(A_i) = A_i$ *and setting* $\pi_i \in \Pi_{n_i}$ *to be* $\pi_i(\vec{v}) = \pi(\vec{v} + \vec{v}_i) - \vec{v}_i$ *we have*

(d) $m(\pi_i) < \delta$ *for all* i;
we then must have that $m(\pi) < \varepsilon$.

The picture of what Axiom C5 requires is that if B_n is almost completely covered by translates of various B_{n_i} each of which is fixed by π and if when π is viewed as a permutation of each translate it is small, then π itself must be small.

Axiom C6. (*Continuity of rearranging large lumps near zero*)
For all $\varepsilon > 0$ *there is an* N *and* $\delta > 0$ *so that if* $N_1 \geq N$ *and* $K_1, K_2, \ldots, K_s \subseteq B_n$ *and*

(a) *the* $K_i = B_{N_1} + \vec{v}_i$ *are disjoint and*

$$\frac{|\cup_{i=1}^{s} K_i|}{|B_n|} > 1 - \delta,$$

(b) *there exist* $\vec{u}_i \in B_n$, $i = 1, \ldots, s$ *so that* $K_i + \vec{u}_i = K_i'$ *are also disjoint and* $\subseteq B_n$,

(c) *the permutations* $\pi, \pi' \in \Pi_n$ *are such that* $\pi(K_i) = K_i$ *and* $\pi'(K_i') = K_i'$ *and for* $\vec{v} \in K_i$

$$\pi'(\vec{v} + \vec{u}_i) = \pi(\vec{v}) + \vec{u}_i,$$

(*i.e.* π *acts on* K_i *exactly as* π' *acts on* K_i') *and,*

(d) $m(\pi) < \delta$

we then must also have $m(\pi') < \varepsilon$.

This last axiom says that smallness of a permutation is a local property. More precisely, if you have a small permutation (in m) that maps a lot of sets K_i to themselves, where the K_i are all translates of a large box B_{N_1}, and if you simply rearrange where these boxes K_i happen to lie in B_n then the permutation you obtain will still be small.

Suppose now that m is a p-size as defined by these six axioms. Suppose α is an arrangement and $\phi \in FG(\mathcal{O})$. Let $x \in X$ and let $f_x^{\alpha,\phi}$ be the associated bijection of $\mathbb{Z}^d$. Recall that for μ-a.e. $x \in X$ for any $\varepsilon > 0$ there is an $N(x, \varepsilon)$ so that for all $n \geq N$ the permutation $\pi_{x,n}^{\alpha,\phi}$ agrees with $f_{x,n}^{\alpha,\phi}$ all but ε of the "time", i.e. on all but a subset of B_n of density at most ε. Define:

$$m_{n,x}(\alpha, \phi) = \min_{\pi \in \Pi_n} \max\{m(\pi), \frac{|\{\vec{v} \in B_n ; \pi_{x,n}^{\alpha,\phi}(\vec{v}) \neq \pi(\vec{v})\}|}{|B_n|}\} \text{ and}$$

$$m_x(\alpha, \phi) = \liminf_{n \to \infty} m_{n,x}(\alpha, \phi).$$

Finally define the p-size m on rearrangements by

$$m(\alpha, \phi) = \operatorname*{ess\,inf}_{x \in X} m_x(\alpha, \phi).$$

Theorem A.2.2. *If* m *is a p-size as defined above, then* m *satisfies Axioms B1–B4 and B5′ and hence the corresponding* m^2 *is an r-size giving the same equivalence classes.*

Proof As the axiomatization given in Axioms B1–B5 was derived from the work in [25] we can just about read off everything we want directly. Axiom B1 is precisely the conclusion of Lemma 2.3 of [25]. Axiom B2 is simply Lemma 2.4 of [25] reworded a bit using the blocks B_n as a Følner sequence. Axioms B3 and B4 are precisely the conclusions of Lemmas 2.5 and 2.6 of [25].

To see that m satisfies Axiom B5′ takes a bit of thought. It is essentially embodied in Axiom C5. To see this first notice that although m is defined as a lim inf in n, Theorem 2.1 of [25] tells us that if this $\liminf_n$ is small then the $\hat{m}(\alpha, \phi) = \limsup_n$ is also small. More precisely, for any δ_1 there is a δ so that if $m(\alpha, \phi) < \delta$ then $\hat{m}(\alpha, \phi) < \delta_1$. This means once n is large enough, for all but $\varepsilon/10$ of the $x \in X$, there is a permutation $\pi(x)$ of B_n so that

(1) $m(\pi(x)) < \delta_1$ and

(2) $\#(\{\vec{v} \in B_n : \pi(x)(\vec{v}) \neq f_x^{\alpha,\phi}(\vec{v}\}) < \varepsilon \# B_n/10$.

Suppose a permutation π of a large box B_N agrees with some π' on all but a fraction δ_1 of B_N, and that we obtain π' by covering all but δ_1 of the B_N by disjoint translates of B_n and permuting the points in each by a permutation of m-size less then δ_1 and applying the identity to the rest of the points. Combining Axioms C1 and C5, if δ_1 is small enough, we will have $m(\pi) < \varepsilon/2$.

Combining the ergodic theorem, the strong Rokhlin lemma and a bit of thought one sees that if (β, ψ) is close in distribution to (α, ϕ) then there will be large blocks B_N in the orbit of a.e. x' where $\pi_{x',N}^{\beta,\psi}$ will be as described as π is described above. We conclude that $m_x(\beta, \psi) < \varepsilon$ and the result follows. □

Bibliography

[1] L. M. Abramov, On the entropy of a flow, *Dokl. Akad. Nauk. SSSR* **128** (1959), 873–875.

[2] W. Ambrose, Representation of ergodic flows, *Ann. of Math.* **42** (1941), 723–739.

[3] W. Ambrose and S. Kakutani, Structure and continuity of measurable flows, *Duke Math. J.* **9** (1942), 25–42.

[4] M. Brin and G. Stuck, Introduction to dynamical systems, unpublished manuscript, contact the authors.

[5] D. L. Cohn, *Measure theory*, Birkhauser, Boston, Mass., 1980.

[6] A. Connes, J. Feldman, and B. Weiss, An amenable equivalence relation is generated by a single transformation, *Ergodic Theory Dynamical Systems* **1** (1981), no. 4, 431–450.

[7] I. P. Cornfeld, S. V. Fomin, and Ya. G. Sinai, *Ergodic theory*, Springer-Verlag, 1982.

[8] A. del Junco, A. Fieldsteel, and D. J. Rudolph, α-equivalence: a refinement of Kakutani equivalence, *Ergodic Theory Dynamical Systems* **14** (1994), no. 1, 69–102.

[9] A. del Junco and D. J. Rudolph, Kakutani equivalence of ergodic $\mathbb{Z}^n$ actions, *Ergodic Theory Dynamical Systems* **4** (1984), no. 1, 89–104.

[10] H. A. Dye, On groups of measure preserving transformations I, *Amer. J. Math.* **81** (1959), 119–159.

[11] J. Feldman, New K-automorphisms and a problem of Kakutani, *Israel J. Math.* **24** (1976), no. 1, 16–38.

[12] A. Fieldsteel, Factor orbit equivalence of compact group extensions, *Israel J. Math.* **38** (1981), no. 4, 289–303.

[13] A. Fieldsteel and N. Friedman, Restricted orbit changes of ergodic $\mathbb{Z}^d$ actions to achieve mixing and completely positive entropy, *Ergodic Theory Dynamical Systems* **6** (1986), no. 4, 505–528.

[14] A. Fieldsteel and D. J. Rudolph, Stability of m-equivalence to the weak Pinsker property, *Ergodic Theory Dynamical Systems* **10** (1990), no. 1, 119–129.

[15] ———, An ergodic transformation with trivial Kakutani centralizer, *Ergodic Theory Dynamical Systems* **12** (1992), no. 3, 459–478.

[16] J. R. Hasfura-Buenaga, The equivalence theorem for $\mathbb{Z}^d$-actions of positive entropy, *Ergodic Theory Dynamical Systems* **12** (1992), no. 4, 725–741.
[17] B. Hasselblatt and A. Katok, *Introduction to the modern theory of dynamical systems*, Encyclopedia of Mathematics and its applications, vol. 54, Cambridge University Press, 1995.
[18] D. Heicklen, Bernoullis are standard when entropy is not an obstruction, *Israel J. Math.* **107** (1998), 141–155.
[19] ———, Entropy and r-equivalence, *Ergodic Theory Dynamical Systems* **18** (1998), no. 5, 1139–1157.
[20] C. Hoffman and D. Rudolph, Uniform endomorphisms which are isomorphic to a Bernoulli shift, preprint submitted to *Annals of Mathematics*.
[21] K. Jacobs, *Measure and integral*, Academic Press, New York–London, 1978.
[22] S. Kakutani, Induced measure preserving transformations, *Proc. Imp. Acad. Tokyo* **19** (1943), 635–641.
[23] J. W. Kammeyer, A classification of the isometric extensions of a multidimensional Bernoulli shift, *Ergodic Theory Dynamical Systems* **12** (1992), no. 2, 267–282.
[24] ———, A classification of the finite extensions of a multidimensional Bernoulli shift, *Trans. Amer. Math. Soc.* **335** (1993), no. 1, 443–457.
[25] J. W. Kammeyer and D. J. Rudolph, Restricted orbit equivalence for ergodic Z^d actions I, *Ergodic Theory Dynamical Systems* **17** (1997), no. 5, 1083–1129.
[26] A. Katok, Monotone equivalence in ergodic theory, *Izv. Akad. Nauk. SSSR Ser. Mat.* (1977), no. 1, 104–157, 231.
[27] ———, The special representation theorem for multi-dimensional group actions, (1977), 117–140, *Astérisque*, No. 49.
[28] A. N. Kolmogorov, A new metric invariant of transient dynamical systems and automorphisms in Lebesgue spaces, *Dokl. Akad. Nauk SSSR* **119** (1958), 861–864.
[29] ———, Entropy per unit time as a metric invariant of automorphisms, *Dokl. Akad. Nauk SSSR* **124** (1959), 754–755.
[30] W. Krieger, On the Araki–Woods asymptotic ratio set and non-singular transformations of a measure space, (1970), 158–177. Lecture Notes in Math., Vol 160.
[31] D. Lind, Locally compact measure preserving flows, *Advances in Math.* **15** (1975), 175–193.
[32] D. S. Ornstein, Bernoulli shifts with the same entropy are isomorphic, *Advances in Math.* **4** (1970), 337–352.
[33] ———, Factors of Bernoulli shifts are Bernoulli shifts, *Advances in Math.* **5** (1970), 349–364.
[34] ———, Two Bernoulli shifts with infinite entropy are isomorphic, *Advances in Math.* **5** (1970), 339–348.
[35] ———, *Ergodic theory, randomness and dynamical systems*, Yale University Press, 1974.
[36] D. S. Ornstein and P. Shields, An uncountable family of K-automorphisms, *Advances in Math.* **10** (1973), 63–88.
[37] D. S. Ornstein and B. Weiss, Entropy and isomorphism theorems for actions of amenable groups, *J. Analyse Math.* **48** (1987), 1–141.

[38] D. S. Ornstein, B. Weiss, and D. J. Rudolph, Equivalence of measure preserving transformations, *Mem. Amer. Math. Soc.* **37** (1982), no. 262.

[39] K. Petersen, *Ergodic theory*, Cambridge University Press, 1983.

[40] S. Polit, *Weakly isomorphic maps need not be isomorphic*, Ph.D. thesis, Stanford Univ., 1974.

[41] V. A. Rohlin, On the fundamental ideas of measure theory, *Amer. Math. Soc. Translation* **1952** (1952), no. 71, 1–54.

[42] A. Rothstein and R. Burton, *Isomorphism theorems in ergodic theory*, unpublished manuscript.

[43] D. Rudolph, Restricted orbit equivalence, *Mem. Amer. Math. Soc.* **54** (1985), no. 323.

[44] ______, *Fundamentals of measurable dynamics*, Oxford University Press, 1990.

[45] D. J. Rudolph, A two-valued step-coding for ergodic flows, *Math. Z.* **150** (1976), no. 3, 201–220.

[46] ______, Classifying the isometric extensions of a Bernoulli shift, *J. Analyse Math.* **34** (1978), 36–60.

[47] ______, *Markov tilings of* $\mathbb{R}^n$ *and representations of* $\mathbb{R}^n$ *actions*, (1989), 271–290.

[48] A. Sahin, Tiling representations of $\mathbb{R}^2$ actions and α-equivalence in two dimensions, *Erg. Th. & Dyn. Sys.* **18** (1998), no. 5, 1211–1255.

[49] E. Sataev, An invariant of monotone equivalence that determinines the factors of automorphisms that are monotonely equivalent to the Bernoulli shift, *Izv. Akad. Nauk SSSR Ser. Mat.* **41** (1977), no. 1, 158–181, 231–232.

[50] P. Shields, *The theory of Bernoulli shifts*, University of Chicago Press, 1973.

[51] Ya. G. Sinai, On the concept of entropy for a dynamical system, *Dokl. Akad. Nauk SSSR* **124** (1959), 768–771.

[52] ______, On a weak isomorphism of transformations with invariant measure, *Mat. Sb.* **63** (1964), 23–42.

[53] A. M. Stepin, The entropy invariant of decreasing sequences of measurable partitions, *Funkcional. Anal. i Priložen* **5** (1971), no. 3, 80–84.

[54] C. E. Sutherland, *Notes on orbit equivalence; Krieger's theorem*, vol. 23, Univ. of Oslo Lecture Notes, 1976.

[55] J.-P. Thouvenot, Quelques propriétés des systèmes dynamiques qui se decomposent en un produit de deux systèmes dont l'un est un schèma de Bernoulli, *Israel J. Math.* **21** (1975), no. 2–3, 177–207.

[56] ______, On the stability of the weak Pinsker property, *Israel J. Math.* **27** (1977), no. 2, 150–162.

[57] V. S. Varadarajan, Measures on topological spaces, *Amer. Math. Soc. Transl., Ser.* 2 **48** (1965), 161–228.

[58] A. Vershik, A theorem on lacunary isomorphism of monotonic sequences of partitionings, *Funkcional. Anal. i Priložen* **2** (1968), no. 3, 17–21.

[59] ______, A continuum of pairwise nonisomorphic dyadic sequences, *Funkcional. Anal. i Priložen* **5** (1971), no. 3, 16–18.

[60] ______, *Theory of decreasing sequences of measurable partitions*, St. Petersburg Math. J. **6** (1995), no. 4, 705–761.

[61] J. von Neuman, Proof of the quasi-ergodic hypothesis, *Proc. Nat. Acad. Sci. USA* **18** (1932), 70–82.

[62] P. Walters, *An introduction to ergodic theory*, Springer-Verlag, New York, 1982.

Index